The Instrument of Science

Roughly, instrumentalism is the view that science is primarily, and should primarily be, an instrument for furthering our practical ends. It has fallen out of favour because historically influential variants of the view, such as logical positivism, suffered from serious defects.

In this book, however, Darrell P. Rowbottom develops a new form of instrumentalism, which is more sophisticated and resilient than its predecessors. This position—'cognitive instrumentalism'—involves three core theses. First, science makes theoretical progress primarily when it furnishes us with more predictive power or understanding concerning observable things. Second, scientific discourse concerning unobservable things should only be taken literally in so far as it involves observable properties or analogies with observable things. Third, scientific claims about unobservable things are probably neither approximately true nor liable to change in such a way as to increase in truthlikeness.

There are examples from science throughout the book, and Rowbottom demonstrates at length how cognitive instrumentalism fits with the development of late nineteenth- and early twentieth-century chemistry and physics, and especially atomic theory. Drawing upon this history, Rowbottom also argues that there is a kind of understanding, empirical understanding, which we can achieve without having true, or even approximately true, representations of unobservable things. In closing the book, he sets forth his view on how the distinction between the observable and unobservable may be drawn, and compares cognitive instrumentalism with key contemporary alternatives such as structural realism, constructive empiricism, and semirealism.

Overall, this book offers a strong defence of instrumentalism that will be of interest to scholars and students working on the debate about realism in philosophy of science.

Darrell P. Rowbottom is Professor of Philosophy at Lingnan University, and serves as Editor in Chief of *Studies in History and Philosophy of Science Part A*. He is the author of *Popper's Critical Rationalism: A Philosophical Investigation* (Routledge, 2011) and *Probability* (2015). He also co-edited *Intuitions* (2014) with Anthony R. Booth. He has published numerous articles in leading journals, including *Analysis, Australasian Journal of Philosophy, British Journal for the Philosophy of Science, Philosophy and Phenomenological Research, Philosophy of Science, Synthese,* and *Studies in History and Philosophy of Science (Parts A and C).*

Routledge Studies in the Philosophy of Science

The Instrument of Science
Scientific Anti-Realism Revitalised

Darrell P. Rowbottom

Routledge
Taylor & Francis Group

NEW YORK AND LONDON

First published 2019
by Routledge
52 Vanderbilt Avenue, New York, NY 10017

and by Routledge
2 Park Square, Milton Park, Abingdon, Oxon OX14 4RN

Routledge is an imprint of the Taylor & Francis Group, an informa business

First issued in paperback 2021

Library of Congress Cataloging-in-Publication Data
Names: Rowbottom, Darrell P., 1975– author.
Title: The instrument of science : scientific anti-realism revitalised / Darrell P. Rowbottom.
Description: 1 [edition]. | New York : Taylor & Francis, 2019. | Series: Routledge studies in the philosophy of science ; 19 | Includes bibliographical references and index.
Identifiers: LCCN 2019005339 | ISBN 9780367077457 (hardback)
Subjects: LCSH: Science—Philosophy. | Instrumentalism (Philosophy) | Realism.
Classification: LCC Q175 .R69 2019 | DDC 501—dc23
LC record available at https://lccn.loc.gov/2019005339

ISBN: 978-0-367-07745-7 (hbk)
ISBN: 978-1-03-209347-5 (pbk)
ISBN: 978-0-429-02251-7 (ebk)

Typeset in Sabon
by Apex CoVantage, LLC

To my colleagues and postgraduate students at Lingnan University, past and present. Thank you for making work such fun.

Contents

Acknowledgements

My work on this book was supported by a Humanities and Social Sciences Prestigious Fellowship from Hong Kong's Research Grants Council, and by a Senior International Research Fellowship from Durham University's Institute of Advanced Study (in conjunction with the European Union via the COFUND scheme). Both grants were appropriately named 'The Instrument of Science'.

I have received useful comments on this work (or parts thereof) from many people, including the following: Sarah Aiston, Haktan Akcin, Jamin Asay, Chris Atkinson, Alan Baker, Peter Baumann, Alexander Bird, Tony Booth, Otávio Bueno, Nancy Cartwright, Anjan Chakravartty, Finnur Dellsén, Michael Devitt, Alex Ehmann, Catherine Elgin, Roman Frigg, Greg Frost-Arnold, Axel Gelfert, Mikkel Gerken, Stephan Hartmann, Robin Hendry, Ian Kidd, Daniel Kodaj, Paisley Livingston, Dan Marshall, Patrick McGivern, Brad Monton, Joe Morrison, Maureen O'Malley, Michael Morris, Bence Nanay, Nikolaj Nottelmann, Wendy Parker, William Peden, Huw Price, Julian Reiss, Samuel Ruhmkorff, Kyle Stanford, Peter Vickers, and Jiji Zhang. I am also grateful for extensive comments from four anonymous referees for Routledge, as well as to anonymous referees of the published papers on which some of the chapters are based (especially for *Studies in History and Philosophy of Science*, *Philosophy of Science*, and *Synthese*).

Science as an Instrument
An Introduction

Broadly, instrumentalism is the view that science is primarily, and should primarily be, an instrument for furthering our practical ends. More precisely, 'instrumentalism' refers to a complex cluster of related views in an empiricist philosophical tradition, in which this book is anchored. Figures well known for their work in this tradition include Ernst Mach, Pierre Duhem, Henri Poincaré, Percy Bridgman, the logical positivists, and Niels Bohr. Less well known is that several nineteenth-century British scientists, such as Kelvin and Lodge, also had an instrumentalist approach to their work. Or so this book later contends.

Instrumentalism has fallen out of fashion largely because historically influential variants thereof suffered from serious philosophical defects, and because it has been misunderstood—as being narrower and more limited than it need be—as a result. The main aim of this book is to challenge this trend by developing a new form of instrumentalism, which is more resilient than its ancestors. I dub this 'cognitive instrumentalism'.

Cognitive instrumentalism has three core components. The first concerns the value of science (or what is central to scientific progress). The second concerns which parts of scientific discourse should be taken literally. And the third concerns what we can reasonably expect science to achieve (or what scientific methods reliably deliver). In brief, these components are as follows: progress in science centrally involves—the value of science lies mainly in—what it enables us to understand about and do with observable things; scientific discourse may only be understood literally when it is grounded in talk about observable things; and science may reliably progress (or increase in value) without discovering new truths (or approximate truths) about unobservable things.

I call the instrumentalism I propose 'cognitive' for two reasons. First, it involves the idea that science is a cognitive tool—a tool for *understanding* the phenomena—rather than just a tool for 'saving' (or predicting) the phenomena.[1] Second, it involves the idea that talk of unobservable things within science is primarily a cognitive tool for comprehending how observable things behave.

The main opposing view to instrumentalism—but far from the only opposing view—is scientific realism.[2] Like instrumentalism, this has several variants and is best understood as referring to a rich philosophical corpus, involving a complicated cluster of related views about science.[3] For the purposes of contrast, however, here is a representative, moderate, version. Moderate scientific realism says: science makes progress primarily when its theoretical content increases in truthlikeness; scientific discourse should *usually* be taken literally, even when it concerns unobservable things possessing properties with which we're not experientially acquainted; well-confirmed theories in mature areas of science are *typically* approximately true (even in what they say about the unobservable); and when the theoretical landscape in mature science changes due to the application of scientific methods, this is *probably* in such a way as to make it more truth-like.

In the course of arguing for cognitive instrumentalism, I also argue against each component of moderate scientific realism. This is a Herculean task, as a result of the intuitive plausibility of moderate realism and the voluminous, ever expanding, literature on the topic (and cognate areas). As a result, I will be delighted if I achieve the modest goal of showing that cognitive instrumentalism is at least as plausible as moderate scientific realism. I also hope to show that it's a reasonable alternative to several 'selective' forms of realism about science, such as entity realism, structural realism, and semirealism.

Before I begin, I'll provide a brief outline of the book.

Structure and Contents: An Outline

As mentioned previously, cognitive instrumentalism has three core components. Each of the first three chapters of the book concentrates on one of these components. Chapter 1 argues that science makes theoretical progress primarily when it furnishes us with more predictive power or understanding concerning observable things. Chapter 2 argues that scientific discourse concerning unobservable things should only be taken literally in so far as it involves observable properties or analogies with observable things. And Chapter 3 argues that what science says about unobservable things is probably neither approximately true nor liable to change in such a way as to increase in truthlikeness (although this is no obstacle to theoretical progress being made in the sense defended in Chapter 1). The central arguments in each of these chapters are independent. I therefore encourage readers to consider the merits of cognitive instrumentalism's discrete constituent parts, in addition to the merits of cognitive instrumentalism as a whole.

The fourth chapter shows how cognitive instrumentalism fits with the development of atomic theory between 1885 and 1930, which is typically taken to be a period in which our knowledge of the unobservable 'realm' increased considerably. *Inter alia*, it illustrates how models that were explicitly taken to be non-literal played a significant role in science of the period, and how such models were championed by British scientists of the time as vehicles for understanding. It also shows how this style of modelling led Bohr to his astonishingly fertile model of the atom. It closes by examining the introduction of 'spin', and considering, with reference back to Chapter 2, to what extent talk of this should be taken literally.

The fifth chapter picks up on the topic of understanding. It develops an account of the form of understanding (why), namely empirical understanding, which was valued by the Victorian scientists discussed in the previous chapter. This form of understanding doesn't require the use of accurate representations (or models). Rather, it involves representations that are cognitively appropriate to serving scientists' empirical ends, in so far, for example, as they are highly memorable and easy to use.

The penultimate chapter deals with two remaining tasks. First, it discusses how the distinction between the observable and the unobservable may be drawn, such that it has epistemic or pragmatic significance, and to what extent unobservable things may become observable due to the development of new instruments or theories. Second, it compares cognitive instrumentalism with three other key alternatives, namely structural realism, constructive empiricism, and semirealism. It also gives some brief arguments for preferring cognitive instrumentalism to these alternatives.

The seventh and final chapter addresses the question of why scientific realism may still seem so appealing, even in the light of well-articulated alternatives such as structural realism and cognitive instrumentalism. It shows there is some evidence, from empirical psychology, that this is as a result of a cognitive bias.

The book finishes with an appendix on how to best characterise scientific realism, and hence the realism debate, about science. I originally intended this to be introductory material. However, the issue requires a much more detailed and lengthy treatment than I anticipated. In the interests of maintaining pace and focus, this material is therefore better in an appendix. This book is primarily about how to think about science, rather than how to think about past thinking about science.

I would like to end with a plea to read carefully and with an open mind. In part, I ask because the dispute concerning realism is often impassioned. When it comes to practical matters, however, what unites scientific realists and most contemporary anti-realists (in the analytic philosophical tradition) is more important than what divides them. Most notably, scientific realists and anti-realists tend to take science to be a form of inquiry *par excellence*, and to take a dim view of attempts to subvert it by 'creation scientists', homoeopathists, and so forth.

I also ask because my task is daunting in so far it requires engagement with many different areas of study. Outside philosophy, I draw on the history of the natural sciences, the natural sciences themselves, and psychology. Inside philosophy, I discuss (formal and informal) epistemology, general philosophy of science, metaphysics, philosophy of language, and philosophy of the natural sciences. I address and connect findings in each of these areas, in developing and defending cognitive instrumentalism. Sadly, however, it is impossible to do full justice to the extensive scholarship relevant to the overarching topic, even within the confines of a monograph.

Notes

1 This goes against the way that instrumentalism is often characterised. For example, Sober (1999: 5) writes that:

> Instrumentalism does not deny that theories are and ought to be judged by their simplicity, their ability to unify disparate phenomena, and so on. However, instrumentalism regards these considerations as relevant only in so far as they reflect on a theory's predictive accuracy. If two theories are predictively equivalent, then a difference in simplicity or unification makes no difference, as far as instrumentalism is concerned.

It is unusual to find authors suggesting instead, as Preston (2003: 261–262) does, that something akin to 'explanatory power' can also play a part in instrumentalism. See Chapter 5 for further discussion.
2 In a recent *philpapers.org* survey concerning philosophical positions, 75% of respondents, 699 out of 931, leaned towards or supported scientific realism. Only 11.6% leaned towards or supported anti-realism. Even out of the 150 professional respondents who listed 'General Philosophy of Science' as an area of specialism, moreover, 56 (37.3%) accepted scientific realism and 25 (16.7%) leaned towards scientific realism. Only 11 (7.3%) accepted scientific anti-realism; 25 (16.7%) leaned towards it.
3 As Hacking (1983: 26) notes: 'Definitions of "scientific realism" merely point the way. It is more an attitude than a clearly stated doctrine. . . . Scientific realism and anti-realism are . . . movements. We can enter their discussions armed with a pair of one-paragraph competing definitions, but once inside we shall encounter any number of competing and divergent opinions'.

1   Scientific Progress and the Value of Science[1]

> If one takes science seriously, then one always considers it also as an instrument. Otherwise, what would it amount to? Building up houses of cards, empty of any application whatsoever!
>
> —Bruno de Finetti (2008: 53)

> Scientific systematization is ultimately aimed at establishing explanatory and predictive order among the bewilderingly complex "data" of our experience, the phenomena that can be directly "observed" by us.
>
> —Carl Hempel (1958: 41)

> The biological task of science is to provide the fully developed human with as perfect a means of orientating himself as possible.
>
> —Ernst Mach (1984: 37)

How does science progress? The boldest answer, which is championed by Bird (2007b, 2008), is by increasing its stock of knowledge. This is bold because 'knowledge' entails justified true belief in the standard epistemological way that Bird intends it.[2] One may refute Bird's view by finding a case in which progress occurs, but no such belief—at the level of the community or any individual therein—is introduced. I have previously argued that there are several such cases (Rowbottom 2008b, 2010a).[3] Sometimes, new true beliefs become stable although they're unjustified (unbeknownst to their possessors). Other times, beliefs of scientists remain unchanged but valuable new information becomes available (e.g. via a computer simulation, or an automated experiment, being completed). On still further occasions, new beliefs are justified and *truer than* their predecessors without being true. In all these cases, progress might occur.

I will not argue against Bird's view of progress on any of these fronts here. Instead, I will focus primarily on the significance of truth, which is also pertinent to a less bold, and correspondingly more plausible, realist account of scientific progress. This is championed by Niiniluoto (2002, 2011, 2014), among others, and is as follows: science makes progress when its content, especially its theoretical content, increases in verisimilitude (or scope and closeness to truth).[4] In essence, Bird thinks that Niiniluoto's view is too broad. But I think that Niiniluoto's view is too narrow (and hence that Bird's is too narrow). More precisely, I will argue as follows. First, scientific progress is possible in the absence of increasing verisimilitude in science's theories (or increasing accuracy in science's models). Second, some central aspects of scientific progress do not involve science's theories increasing in verisimilitude. Third, increasing predictive power and understanding of how phenomena interrelate is most central to how science progresses. Fourth, increasing predictive power may involve increasing scientific 'know how' (as opposed to theoretical knowledge).

Many philosophers have discussed these matters in terms of 'the aim of science'. They tend to think as follows. Progress is made by achieving, or at any rate getting closer to achieving, the aim. I prefer to resist thinking in this way, for reasons I explain in the Appendix. Here, I'll put my case tersely. First, it's a clumsy metaphor. Science doesn't have aims. People do. Second, it's a confusing metaphor. The so-called aim of science doesn't have anything to do with the aims of scientists, or any other people. Hence, I prefer to be explicit in advancing an *evaluative* or *normative* thesis about science, as I will shortly make clear.[5]

1. Scientific Progress Without Increasing Verisimilitude

How can we show that scientific progress should not be 'defined by increasing verisimilitude' (Niiniluoto 2014: 77)? We need only find a single case, hypothetical or actual, where the verisimilitude of scientific theories fails to increase, yet science nevertheless makes progress. Then we may conclude that increasing verisimilitude in the content of science is *not necessary for*, even if it turns out to be sufficient for, scientific progress to be made.

I'll give such a hypothetical case in the next paragraph. But beforehand, I must prepare the ground by signalling my agreement with Niiniluoto (2011) on two matters. First, the notion of scientific progress is normative: 'the theory of scientific progress is not merely a descriptive account of the patterns of developments that science has in fact followed. Rather, it should give a specification of the values or aims that can be used as the constitutive criteria for "good science".' Hence, '[t]he task of finding and defending such standards is a genuinely philosophical one which can be enlightened by history and sociology but which cannot be reduced to empirical studies of science' (ibid.). Second, progress may be multi-faceted: 'Progress is a goal-relative concept. But even when we consider science as a knowledge-seeking cognitive enterprise, there is no reason to assume that the goal of science is one-dimensional' (ibid.). This is the case on Niiniluoto's own account, in so far as both truth *and* informativeness (or what Kuhn called 'scope') are significant. On a one-dimensional account focused on truth, by way of contrast, making modern science's claims closer to the truth by limiting its domain—e.g. by discarding all scientific theories except those of biomechanics—could be progressive.

Now consider the following thought experiment.[6] Imagine that all the leading scientists working in a specific area of physics, such as mechanics, have gathered to discuss the state of their field. They unanimously agree that they have identified the true general theory in the domain. Moreover, it is *true* that they have identified the true general theory in the domain, and their beliefs that they have done so are *justified*. (So estimated verisimilitude equals actual verisimilitude.) They discuss what they should do next. Is any further scientific progress possible in their area? If not, then it would be reasonable for them to crack open a jeroboam of champagne, celebrate their great success, and move on to new things.

Niiniluoto's view entails that no more scientific progress is possible in the area, because a true *general* (i.e., maximally verisimilar) theory has been found. The corollary is that it would be acceptable for the scientists to cease working in that area.[7] But this is incorrect. Why? First, the true theory could be difficult, or even impossible, to use for predictive purposes. It could concern initial conditions that are beyond our ability to measure and determine: it could be practically impossible to find reliable values for some of the variables in the equations. Or the theory could merely be unusable in many situations in which timely predictions are desirable, due to the need for arduous calculations.

Second, even if it were of considerable predictive use, the theory might fail to measure up to our explanatory expectations. Imagine that it involved considering numerous variables, such that it was hard to appreciate how changes in one would tend to affect changes in another, in many applications, without the use of extensive computer simulations. It would not serve to grant insight, or to build 'physical intuition'. Consider also the role of models, and approximations and idealisations, in bringing theories into contact with experience. Think of pendulum motion. Knowing the true theory thereof is simply not enough. We want to know the models, like that of the simple pendulum, too. For example, you could know the theory of real pendulum motion, in principle, without spotting that when the angle of swing is small, the sine of the angle is approximately equal to the angle, and hence that the motion is approximately simple harmonic provided that (approximately) no damping occurs. That's because the complete true theory is rather complex; it includes factors dealing with friction at the bearing and the mass of the rod, refers to the sine of the angle, and so on. It obscures the result that pendulum motion is sometimes approximately simple harmonic. It is true that this result *can* be derived from the true theory. But such derivations are sometimes extremely difficult, and making them requires a lot of effort.

The previous two paragraphs make the following key claim: less true theories or models might be superior to their more true counterparts when it comes to helping us to predict and to understand (how observable things behave). The following sub-section provides a case study in support.

1.1 The Pendulum—A Case Study on Accuracy, Simplicity, and Scope

Consider how the equation of motion governing the simple pendulum is derived. This begins with numerous idealisations, such as:

(i)   The pendulum only moves in two dimensions.
(ii)   The rod holding the bob has no mass.
(iii) The rod holding the bob is perfectly rigid.
(iv)   The pendulum is moving in a vacuum (or experiences no frictional forces from movement through the fluid in which it is placed).
(v)   The pendulum is in a uniform gravitational field.
(vi)   There is no friction at the pivot from which it swings.
(vii) The bob has uniform density.[8]

Let the distance between the pivot and the centre of mass of the bob be *l*, and the mass of the bob be *m*. Let the gravitational field strength be *g*, and the angular displacement of the rod be θ.

As shown in Figure 1.1, the equation for the restorative force acting on the bob perpendicular to the rod, *F*, is as follows:

$$F = -mg\sin\theta \tag{1}$$

(I employ a negative sign because the force acts in opposition to the angular displacement.) And we know also the second law of motion:

$$F_{net} = ma \tag{2}$$

Thus we can derive, by substituting (1) into (2):

$$-g\sin\theta = a \tag{3}$$

Now note that the bob will trace the arc of a circle, of radius *l*. The equation governing the length of the arc traced, *s*, relative to the angular displacement, is this:

$$s = l\theta \tag{4}$$

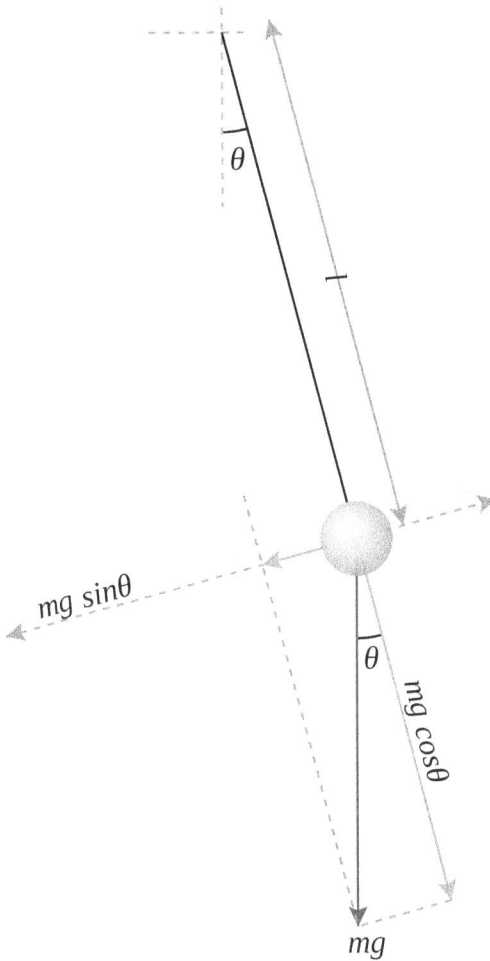

Figure 1.1 The Simple Pendulum[9]

And by taking the second derivative of this, with respect to time, we arrive at another formula for the acceleration (perpendicular to the rod) at any point of the swing:

$$a = \frac{d^2 s}{dt^2} = l\frac{d^2 \theta}{dt^2} \tag{5}$$

Now we may substitute (5) into (3), in order to arrive at the equation for motion:

$$-g\sin\theta = l\frac{d^2 \theta}{dt^2} \tag{6}$$

At this point, we may introduce an approximation (rather than an idealisation):

(viii) The sine of the angle is approximately equal to the angle, when the angle of swing is small.

Thus, for small values of θ,

$$-\frac{g}{l}\theta = \frac{d^2\theta}{dt^2} \tag{7}$$

The solution to (7) shows that the motion is simple harmonic. This is as follows, where θ_{max} represents the maximum angular displacement:

$$\theta(t) = \theta_{max} \cos\left(\sqrt{\frac{g}{l}}t\right) \tag{8}$$

Now let's take some steps towards eliminating any idealisations and approximations. The sole approximation, (viii), is easily removed; let's allow the angle of swing to be large (or demand further precision even when the angle of swing is small). We can now begin again with equation (1), but bear in mind that it only represents *one* component of force relevant to determining the pendulum's motion, and introduce some of the other forces that will be present (for a real pendulum on Earth). First, let's remove idealisation (iv) and introduce drag. This depends on several factors, namely the speed of the pendulum, the shape of the pendulum, and the density of the medium it moves in. More precisely, the standard equation—which is highly similar to one originally proposed by Lord Rayleigh—is as follows:

$$F_{drag} = \frac{1}{2}\rho v^2 CA \tag{9}$$

The density of the medium the pendulum moves through is ρ, and the area on which the medium acts as the pendulum moves is A. (In this case, we can define A as the area of the orthographic projection of the pendulum, both rod and bob, on a plane perpendicular to the velocity vector.) The velocity of the pendulum is v. Finally, C is a coefficient related to the overall geometry of the object.

Now we can work out the velocity with respect to angular displacement by taking the first derivative of equation (4) with respect to time:

$$v = \frac{ds}{dt} = l\frac{d\theta}{dt} \tag{10}$$

And from (9) and (10), using substitution, we arrive at the following:

$$F_{drag} = \frac{1}{2}\rho CAl^2\left(\frac{d\theta}{dt}\right)^2 \tag{11}$$

Let's next endeavour to remove idealisation (vi) by considering friction at the pivot. This will be of two forms; static when the pendulum is still (i.e., at the 'high points' of each swing), and dynamic when it is in motion. We may use the following standard equations:

$$F_{\text{stat}} = \mu_{\text{stat}} F_{\text{n}} \tag{12}$$

$$F_{\text{dyn}} = \mu_{\text{dyn}} F_{\text{n}} \tag{13}$$

Here, μ_{stat} and μ_{dyn} are coefficients of (static and dynamic) friction (which are partially determined by the materials used in, and the design of, the pivot), and F_{n} is the normal force between the surfaces at the pivot. We know that F_{stat} is active only when the pendulum is still, and that F_{dyn} is active only when the pendulum is moving. So we may introduce a function, μ_{θ}, which has the value of μ_{dyn} when $\dfrac{d\theta}{dt}$ is non-zero, and the value of μ_{stat} otherwise. We may then represent the frictional force at the pivot as follows:

$$F_{\text{pivot}} = \mu_{\theta} F_{\text{n}} \tag{14}$$

Now let's assume that the normal force is equal to the component of gravitational force acting downwards on the pendulum:

$$F_{\text{n}} = mg \cos\theta \tag{15}$$

From (14) and (15), we find:

$$F_{\text{pivot}} = \mu_{\theta} mg \cos\theta \tag{16}$$

When we combine equations (1), (11), and (16), we find:

$$F_{\text{net}} = \frac{1}{2}\rho CAl^2 \left(\frac{d\theta}{dt}\right)^2 + mg\left(\mu_{\theta}\cos\theta - \sin\theta\right) \tag{17}$$

Then by substitution from (2), and then (5), into (17):

$$l\frac{d^2\theta}{dt^2} = \frac{\rho CAl^2}{2m}\left(\frac{d\theta}{dt}\right)^2 + g\left(\mu_{\theta}\cos\theta - \sin\theta\right) \tag{18}$$

I might continue in the same vein, but this would only be for effect. So let's assume, for the sake of argument, that there are real pendulums for which this equation is apt *in so far* as idealisations (i), (ii), (iii), (v), and (vii) hold. (I could remove some of those idealisations, e.g. by considering moments of inertia, treating the rod as spring-like, and so forth. But we would only end up at a similar juncture, with a more complicated equation.) Aren't we now *extremely close* to having the true equation governing their motion?

The answer lies in the negative. First, consider the equation for drag (9). The derivation of this is also based on idealisations. Rayleigh (whose family name was Strutt) identified two in his classic paper:

> [M]otions involving a surface of separation are unstable. . . . If from any cause a slight swelling occurs at any point of the surface, an increase of pressure ensues tending not to correct but to augment the irregularity. . . . The formulae proposed in the present paper are also liable to a certain amount of modification from friction which it would be difficult to estimate beforehand.
>
> (Strutt 1964: 294)

Indeed, as Anderson (1997: 104–106) explains, Rayleigh was content with *saving the drag phenomena* via crude, but effective, numerical means:

> [T]he Newtonian values do not even come close to agreeing with the experimental data, but neither do the values obtained by Rayleigh. . . . That did not phase Rayleigh. He simply *adjusted* his findings to agree . . . To accomplish that, he had to multiply his finding by a factor of 2.27. . . . Rayleigh made no apology for such an adjustment. . . . Rayleigh's model of the flow over a flat plate at some angle of attack, with surfaces of discontinuity emanating from the leading and trailing edges . . . was a good idea—it reflected the real, separated flow that actually occurs in nature. His error was in assuming free-stream pressure in that separated region. Today we know that the pressure in that region is less than the free-stream pressure. Because of that error, Rayleigh's theory did not directly advance the state of the art of prediction of aerodynamic drag and therefore was not used to any extent by designers of flying machines in subsequent years. From that point of view, the flow model using surfaces of discontinuity was a blind alley for drag predictions. On the other hand, his theoretical value for the mean pressure over the bottom of the flat plate was reasonably correct . . . [and] is still used in the modern literature.[10]

Even (again) imagining that the drag equation were true, moreover, *it involves variables that depend for their values on further variables that are not mentioned in the equation*. The density of the medium, ρ, depends on the temperature and pressure even if we assume we're dealing with an ideal gas. For air, for instance, it will also depend on the humidity. And all these factors vary over time when any real pendulum is concerned.

Second, consider the equation for friction at the pivot (14). In reality, μ_{dyn} is not a constant; it varies with velocity, and other factors (such as humidity of air, if air be the medium through which the pendulum moves).

It is a *system* variable. It does not depend merely on the properties of the materials and their interface. Moreover, it can vary in extremely complex ways.

Third, note that other variables in equation (18) may change in ways we haven't begun to treat. Imagine the rod and bob are made of iron. As the pendulum swings, the friction will heat it. And this will cause it to expand. Thus A and l will change over time. If the rod and bob are instead made of wood, then the moisture content of the wood will also affect A and l. If the wood is knotty too, the moisture content may also affect C. And so forth.

I am not on a skeptic's flight of fantasy, disconnected from experimental reality. Effects of some of the features I've mentioned can be observed with simple setups. For example, as Peters (2005: 19) illustrates with a straightforward experiment using a solder pendulum, which violates idealisation (iii) as it is not elastic:

> Because it is associated with hysteresis that ultimately involves temperature, through diffusion processes, pendulum motion is extremely complex. . . . [M]any of the complex features can be observed with a pendulum and sensor that are easily built. In fact, some features can be studied . . . simply through visual observation.

Indeed, it is hard to imagine the complexity of the true equation—or the truest equation we could arrive at, in principle—in terms of the variables mentioned so far. Nobody knows it. Peters (2005: 31) even goes so far as to declare: 'The primary challenge to any meaningful attempt to model the pendulum's complex motion derives from the multiplicity of interactions that are obviously present. . . . [P]*hysics is woefully lacking in any of the tools . . . necessary to explain the observations.*' [emphasis added]. That's to say, we aren't even in a position to save the phenomena concerning physical pendulums.[11]

Fourth, and finally, we have been employing macroscopic variables, such as temperature, which can be reduced to more fundamental variables, such as the mean translational speed of molecules. So if we take statistical mechanics seriously, we will need to introduce probabilities in the true equation.[12] Admittedly, such probabilities may only be needed because of our ignorance—our inability, that is to say, to measure the initial conditions in the situations under consideration. Yet suppose that this ignorance is remediable, and that we will one day be able to measure the positions, masses, velocities, etc., of all the particles in such situations. We would still lose any useful generality by considering the equations governing any *specific* pendulum system. We would have a different equation for the motion of each pendulum.

By increasing accuracy we lose simplicity and scope. And by making our model less truth-like, we may increase simplicity and scope at the cost of some accuracy.[13] (And thus we may also, I will argue in due course, improve understanding.) Good scientific modelling consists in appropriately balancing such factors. It is about finding *sweet spots*. This point is not limited to physics. As Friedman (1953: 14) memorably puts it:

> Truly important and significant hypotheses will be found to have "assumptions" that are wildly inaccurate descriptive representations of reality, and, in general, the more significant the theory, the more unrealistic the assumptions (in this sense).[14]

Many historical studies of the pendulum were focused on the aim of improving clocks. So a central aim was to try to create real pendulum systems where the idealisations listed earlier—and others subsequently alluded to—were close to holding. The 'direction of fit' was reality to fantasy.[15] For example, seasonal changes in ambient temperature were responsible for large time-keeping errors in early pendulum clocks. This led to the invention of several ingenious devices, such as the mercury pendulum and the gridiron pendulum. The former, devised in 1721 by George Graham, uses a bob filled with mercury. As the temperature increases (or decreases), the mercury expands (or contracts), and moves the centre of mass of the pendulum upwards (or downwards). This off-sets the downward (or upward) movement of the centre of mass due to the expansion (or contraction) of the rod. The latter, devised in 1726 by John Harrison, employs rods of metals with different thermal expansion coefficients, such as zinc and steel. These are arranged such that expansion (or contraction) of one type of rod tends to decrease (or increase) the length of the pendulum, whereas the expansion (or contraction) of the other type of red tends to increase (or decrease) the length of the pendulum. A balance may then be struck, so that the length of the pendulum remains (approximately) the same when temperature changes occur. Naturally, neither kind of device is perfect. A problem with the former arrangement is that the mercury tends to change in temperature considerably more slowly than the rod. A problem with the latter is that zinc creeps considerably.

1.2 Back to the Thought Experiment

I have made the case that scientific progress is possible in the absence of increasing theoretical verisimilitude. To distil: having true theories doesn't entail having predictive ability, especially in so far as it doesn't entail having (cognitively and practically) useful models; having models that faithfully represent their targets doesn't entail having useful models; and useful models typically don't faithfully represent their targets. But how might my opponents respond? One option is to say that advances such as those discussed previously are in *engineering*, rather than science. My rejoinder is as follows.

First, this inappropriately marginalises applied science. Do we want to say that any work to generate new predictions from a true theory counts as non-scientific? Kuhn, for one, would have disagreed; classification and prediction are both parts of his 'normal science', as detailed in Chapter 3 of *The Structure of Scientific Revolutions*.[16] Do we want to say that Lagrangian mechanics was a development in engineering, rather than physics, in so far as it did not question Newton's laws of motion? It is a curiosity, then, that Lagrangian mechanics is widely regarded as a theoretical development in physics.

Second, more trenchantly, even if we accept that improving the predictive power of theories—e.g. by reformulating them or developing models to apply them—is *not* a scientific activity, such developments promote understanding and insight too. So if we took the 'only engineering remains' line, we would be forced to conclude that improving understanding of how phenomena interrelate does not, by itself, count as scientifically progressive. Crucial, here, is that I take the previous discussion of the pendulum to support not only an account of understanding on which the propositions expressing an understanding need not be true (as defended by Elgin [2007]), but also the view that increases in understanding are often accompanied by decreases in the representational accuracy of a model (as argued by Rancourt [Forthcoming]).[17] That is, in so far as I take us to have considerable understanding of real pendulums, and not just fantasy pendulums.

I will return to this topic in the next chapter, and again in the fourth and fifth. For now, here's a concrete example of the kind of understanding I have in mind. Consider again the simple pendulum, as represented in equation (8), and compare this with all real pendulums of simple construction. None behaves exactly as mathematically represented, when the representation is understood in a fine-grained way. And predictions from the equation will be highly inaccurate, in some respects, for all such real pendulums. For one thing, their motions are all damped; hence, as time goes on, the prediction for the maximum amplitude becomes increasingly worse in any given experiment. (The equation predicts that such pendulums will swing forever.) But note, as Sober (1975: 67) emphasises, that 'Mathematical equations do not merely enable us to calculate the value of one variable given the values of the others. They also characterize the relationships between variables and enable us to compute how any variable changes relative to the others.' And in this case, the relationship between length and frequency of swing *is successfully characterised* by the equation, albeit in a coarse-grained way, for all the aforementioned pendulums. Increasing length decreases (average) frequency of swing, whereas decreasing length increases (average) frequency of swing, over the period in which the pendulums move. The hypothetical scenario lets us see why this is so, and formulate a principle—not *the* principle—behind this. This builds what physicists sometimes call 'physical intuition'.

So the advances I've discussed are not merely in engineering. However, another response to my argument that scientific progress is possible in the absence of increasing theoretical verisimilitude is to deny that the scenario involved in the thought experiment is possible. The most natural way to do this is to suggest that any true general theory, in a given area, must also be maximally virtuous relative to possible alternatives: i.e., maximally simple, comprehensible, predictively useful, and so forth.[18] There is plausibly no empirical evidence to this effect, from history, in so far as we haven't acquired any true general theories. But even accepting that we *have* identified some such theories, the claim is *comparative*. So what empirical grounds do we have for making a conclusion about all *possible* alternatives, even restricting 'possible' to 'conceivable', when we do not know how many unconceived alternatives there are, let alone what they look like? I will return to this 'problem of unconceived alternatives' in Chapter 3.

Moreover, there is considerable evidence from the history of science—similar, in character, to the pendulum example—that theory T (or model M) may be less virtuous than T^* (or M^*) in key respects, even when T (or M) is more verisimilar than T^* (or M^*). For example, the two-spheres model of the universe—where Earth is stationary at the centre, and the sphere of stars rotates around it—is much simpler and easier to use, for navigation at night on the sea, than any truer model that I am aware of. Similarly, it is much easier to make everyday predictions in mechanics by disregarding relativistic considerations—and even by positing non-existent forces, such as the centrifugal, Euler, and Coriolis forces—than it is to take them into account. So why should we expect such results *not* to occur in the special case where T happens to be true, as well as more verisimilar than T^*? It's more plausible to say as follows. Truth is sometimes stranger—less accommodating, less elegant, less simple, less comprehensible—than fiction.

Incidentally, Niiniluoto (2002: 183–184) agrees with this statement—and, moreover, that there is no a priori case to the contrary—in so far as *simplicity* is concerned:

> [S]*implex sigillum veri*. This is a highly dubious metaphysical doctrine: why on earth should nature be simple? Why should there be any a priori reasons to regard simpler theories as more probable than their rivals?

We should conclude that acquiring theories with greater verisimilitude is not necessary for progress, even if it is sufficient for progress.

2. Increasing Verisimilitude vs. Increasing Predictive Power and Understanding of the Phenomena

Niiniluoto (2011) and I agree, recall, that 'there is no reason to assume that the goal of science is one-dimensional'. Thus, one might hold that increasing verisimilitude is the *primary (or central)* dimension of scientific progress, while accepting that there are other unrelated dimensions in which it can progress (even when maximal verisimilitude is achieved). And one might respond to the scenario described in the previous section by judging that further progress could be made, while denying that such forms of progress are as important, or valuable, as increasing verisimilitude (when that's also possible).[19] So, in summary, to grant that increasing verisimilitude is *unnecessary* for making scientific progress is not to deny that increasing verisimilitude is the primary means of making scientific progress.

My response to this subtle move is to proffer another thought experiment, and elicit a judgement on the proper conduct of the scientists therein. Imagine that all the experts in a specific area of chemistry gather together to discuss how to direct future research. They unanimously agree—and are correct (in virtue of knowing, or believing justifiably, if you so wish)—that their two best strategic options will lead to two mutually exclusive outcomes. If they take the first, they will maximise their predictive power concerning, and understanding of, the phenomena with which their branch of chemistry is concerned. If they opt for the second, they will discover the true unified theory in that domain. Their dilemma arises because their resources are limited. Pursuing all the goals simultaneously would result in extremely limited progress in achieving each. Unfortunately, moreover, pursuing the truth will result in limited incidental progress toward maximising predictive power and understanding, and vice versa.

After a heated discussion, the scientists unanimously agree that they should forego the chance to find the true theory of everything in their domain. Instead, they opt to pursue greater predictive power and understanding. Have they made the wrong decision, in some significant sense? Have they failed to do what ideal scientists would do, in such circumstances?

I contend that it is reasonable to answer these questions in the negative, even if it is assumed that the scientists have no pressing practical need to achieve greater predictive power or understanding. And note that this is not to presume that the scientists take the *correct* course of action, in pursuing predictive power and understanding. It suffices for each option to be as good, as scientific, as the other.

Why do I approve of their decision, and think that it is at least as good as the alternative? Grant, for the sake of argument, that finding the truth (about the unobservable, *inter alia*) has considerable intrinsic value.[20] First, why deny that achieving understanding of how phenomena interrelate has similar value?[21] I expect that many realists would *not* deny this, but would instead insist that truth is a necessary condition for understanding. Niiniluoto appears to be such a realist, judging by the following passage (if we assume that understanding and explanation are appropriately related, e.g. that one understands some event or regularity in phenomena only if one can explain it):[22]

> The realist of course appreciates empirical success like the empiricist. . . . But for the realist, the truth of a theory is a precondition for the adequacy of scientific explanations.
>
> (Niiniluoto 2002: 167)

As I argued in the previous section, however, understanding is non-factive. If you need any more convincing, consider the consequences of denying this. One would have to accept that Newton did not understand mechanical phenomena, *except in so far as he understood his own theory of mechanics*, because the predictions of his theory (when conjoined with true statements of initial conditions) are, strictly speaking, false. Consider projectile motion on Earth, for illustrative purposes. Newton was able to predict that the best angle at which to throw a javelin is 45 degrees (if one is throwing it on the flat and wants to maximise its horizontal displacement). But on a factive account of understanding, we should say that he did not *understand* (and/or could not explain) why this is so. That is, provided we accept that relativity theory is more accurate than its classical predecessors. In short:

$$ma \neq \frac{1}{\sqrt{1 - \dfrac{v^2}{c^2}}} ma \text{ when } v \neq 0$$

I do not want to argue over the proper definition of 'explanation'. If anyone wants to insist that explanation requires truth, I will grant it. But I will then deny that such 'explanation' is of central importance in science. What's important is a surrogate, for which I'll keep the label 'understanding'. As I have already indicated, I take us to understand something about real pendulums in virtue of considering fantasy pendulums, possessing properties and exhibiting behaviour that real pendulums never could, for instance. (This is due to special relations that obtain between the real pendulums and the fantasy pendulums. I'll say more on this topic in my discussion of analogies in the next chapter.) I have also argued that this does not require approximate truth, any more than truth *simpliciter*. That's to say, 'understanding requires approximate truth' is a response

I've already blocked. I need not, and do not, deny that understanding requires elements of truth.

But I might even grant at this juncture, for the sake of argument, that understanding is factive. Why then deny that achieving great predictive power has value *equal* to that of finding true theories (with high scope)?[23] After all, acquiring said power involves acquiring a means by which to *derive many new truths* concerning observables. And as I have already shown in the previous section, acquiring the true theory in a domain does *not* entail acquiring the theory that will be the most predictively useful, or indeed a theory that will be of any predictive use whatsoever, in said domain. So I will rest my case on this issue.

I take the following to be a reasonable stance. Achieving maximal verisimilitude is *no more central* to scientific progress than achieving maximal predictive power and understanding. Or to put it differently, the *primary* aspect of scientific progress should not be 'defined [purely] by increasing verisimilitude' (Niiniluoto 2014: 77).

3. Saving and Understanding the Phenomena as Central

So far, I have argued that science progresses neither uniquely when, nor centrally only when, its theories increase in verisimilitude. I will now argue that scientific progress *more centrally* involves increasing our resources for predicting and understanding how phenomena interrelate. I'll use a final thought experiment in order to do so.

WWIII was the war to end all wars. Earth was ravaged by nuclear weapons and only a few outposts of humanity remained. Humans lived underground, for the most part, for several generations. But they have re-emerged, and begun to rebuild, now that the surface is considered safe. Technology is limited; there are no motor vehicles, medicines (other than herbs), electrical appliances, or even generators. Moreover, most of humanity's knowledge, save what could be passed down orally, has been lost. In effect, the world has returned to a pre-industrial state. However, there are stable societies with sufficient resources to support academics. The capital city of one such society, built on the ruins of London, is Novum Londinium. The King has set up a college there, which is called King's College Londinii.

One day some explorers make a great discovery. They find a buried secondary school science library, which has a few textbooks intact. There is no doubt that these books contain precious pre-apocalyptic lore, which was the key to some of the fantastic feats performed by the ancestors, such as curing cancer and artificially creating humans with particular characteristics. There are around ten books in total. Some cover atomic theory—at a high level of abstraction, and without any discussion of historical theories—at first-year undergraduate level. Others cover mathematics to final-year undergraduate level. Further searches are conducted after this discovery, but these are unsuccessful.

The books are presented to the leading natural philosophers of the college, and they hold a grand meeting to decide what the future course of their research should be. They quickly agree that it will prove important to master the mathematics in order to unlock the secrets of the ancestors. They set to work.

Several years pass. They reconvene. They've mastered the mathematics. They now discuss what to do in light of the atomic theory described in the remainder of the books. They understand the basic idea behind it pretty well, they think. It says that everyday materials are made up of atoms, which in turn contain entities called 'electrons', 'neutrons', and 'protons'. It adds that these things all have different properties, e.g. sizes. There are also parts of some of these parts, which are mentioned by name in the available texts, e.g. 'quarks', but are not described any further. There is also a tantalising mention of a grand theory that underlies the atomic theory, which is known as string theory. But this too has been lost.

It's also clear that the atomic theory was once closely connected to experience, and especially to predicting what happens when different kinds of materials are combined. There is mention of a lost art known as 'chemistry', which relied on a 'periodic table' of 'elements'. Some of these elements are mentioned in the text, and correspond to names still in use, like 'gold' and 'silver', although the table itself is not present.

The 'scientists'—for scientists they now take themselves to be, given their findings—debate what to do. They could go in one of two directions. They could either try to work out the more fundamental theories underlying the atomic theory—inquire into what these 'quarks' that compose neutrons and protons might be, and so forth—or try to connect the atomic theory with experience by deriving some observational consequences from it. As their resources are severely limited, they hold that they should devote all their effort to just one avenue. If they attempted both, it's doubtful any significant progress would be made along either.

You will guess what I think it would be natural, and better, for the 'scientists' to do. Indeed, I would even go so far as to suggest that some of the content of the atomic theory might not be properly understood *unless* it could be appropriately linked with experience—the discussion of charges, for example, might be grasped only incompletely without efforts to understand how it related to observable phenomena, such as lightning and ferromagnetic materials. This appears to be reasonable on either an empiricist or a pragmatist view on meaning, although I shall not argue for either here, as I discuss both in my advancing my argument for 'property instrumentalism' in the next chapter.

So let's grant, for present purposes, that the scientists understand (much of) the atomic theory. Now imagine, further, that it is clear to the scientists, from the textbook, that discovering the more fundamental nature of things will not help them to predict or accommodate any phenomena that they cannot already (given the technological limitations of the society—e.g. the lack of particle accelerators). Should the scientists probe into the more fundamental nature of things? It seems *permissible* for them to do so, at the bare minimum, on the view that scientific progress consists in (or centrally involves) finding theories of increased verisimilitude. Going the other way, after all, might only involve making models, and gathering data, to *connect* (or better connect) atomic theory with experience.

My view, however, is that by probing into the more fundamental nature of things they would be electing to do metaphysics rather than science.[24] As Ruse (1982: 72) puts it: 'In looking for defining features, the obvious place to start is with science's most striking aspect—it is an empirical enterprise about the real world of sensation.'

As a result, it might appear that I want to label string theorists—or workers in some areas of theoretical physics—as non-scientists. (String theory allows for an extremely large number of possible manifolds, only one of which obtains.)[25] I confess to an inclination to bite the bullet—to say that this is metaphysics—following Ginsparg and Glashow (1986: 7):

> [Y]ears of intense effort by dozens of the best and the brightest have yielded not one verifiable prediction, nor should any soon be expected. . . . In lieu of the traditional confrontation between theory and experiment, superstring theorists pursue an inner harmony where elegance, uniqueness and beauty define truth . . . Contemplation of superstrings may evolve into an activity . . . to be conducted at schools of divinity by future equivalents of medieval theologians. For the first time since the Dark Ages, we can see how our noble search may end, with faith replacing science once again.[26]

I need not bite the bullet, however, because the situation in the post-WWIII thought experiment is significantly different from the actual situation, when one thinks of science *as a whole*. For only a small percentage of scientists are string theorists, and it is clear that science *as a whole* is intimately connected with the world of sensation. Moreover, although it appears that the prospects for connecting string theory with experimental science are dim, a breakthrough *might* nevertheless happen because it is understood, in the scientific community, how to connect claims from particle physics with experience. So, in summary, in the thought experiment the *whole of* 'science' would lack predictive power, despite involving a quest for the truth, if the inquirers elected to seek more fundamental theories.

4. Increasing Know How

In closing, I should like to register my sympathy with the recent argument of Mizrahi (2013a), to the effect that increasing our *know how*—which, like Ryle, I take to be distinct from propositional knowledge[27]—is another significant means by which science can progress.[28] As Mizrahi points out, echoing Hacking (1983), philosophers of science have had rather a tendency to focus on the *theoretical* aspects of science—on its semantic content—to the neglect of its practical aspects, in a manner that I have already criticised in my mention of applied science in Section 1 of this chapter.[29]

Baird and Faust (1990: 147), whom Mizrahi cites approvingly, put the matter as follows:

> According to most philosophers, experiments are run in order to pro-mote theory; improvements in the ability to experiment are merely instrumental goods that promote the final good of justifying the assertions of a wider and wider domain of sentences. Thus, accord-ing to most philosophers, improved theories account for the progress of scientific knowledge. It is our contention that this asymmetry is a mistake. Technicians, engineers and experimenters, in a vast number of instances, are able to make devices work with reliability and sub-tlety when they can say very little true, or approximately true, about how their devices work. Only blind bias would say that such scien-tists do not *know anything* about nature. Their knowledge consists in the ability to *do* things with nature, not *say* things about nature.

Baird and Faust (1990: n.2) also claim that 'one of the motivations for experimental work is to provide data with which to theorize'. This might seem implausible because scientific publications suggest otherwise (in some disciplines, at least), if they are taken at face value. However, some of my previous work—Rowbottom and Alexander (2011)—illustrates the difference between presentation and practice. Initially, we established that biomechanists privately confessed to going on data 'fishing expedi-tions'.[30] We then independently examined 100 papers on biomechanical topics (published in the leading journals in the area, namely *The Jour-nal of Experimental Biology* and *Journal of Biomechanics*). We found that *no* papers stated exploration as an aim of the experiments presented therein, which indicates the presence of a strong bias against so doing, although 58% of papers had hypothesis testing as a stated aim. Remark-ably, furthermore, we shared strong suspicions (at the bare minimum) that presentational hypotheses—or what one might describe, less charita-bly, as 'fake hypotheses'—were present in 31% of the papers in this latter group.[31] In the majority of these cases, we noted:

> one of the stated aims of the paper was to test a hypothesis that was widely accepted to be true or false—so widely, indeed, that a paper which stated simply that it would test the hypothesis would never have been considered interesting enough to be published unless it ran

counter to the expectation—when it appears that the actual point of the exercise was to gather data in order to understand some phenomenon (or phenomena) better. A sub-set of these cases were hypotheses that we thought to be post hoc.

(Rowbottom and Alexander 2011: 258)

Another way of putting this matter is that science is legitimately concerned with problem solving, and directly trying to grasp how phenomena interrelate, rather than merely testing theories.[32] Moreover, it is not passive. We often want to know how to *intervene* in order to affect what happens; i.e., to predict the effects of our possible actions (as well as to identify new possible actions).

There are two points to take away from this section. First, my use of 'predict and understand how the phenomena interrelate' covers predictions concerning *active interventions* in order to shape what happens. Second, the development of new instruments can serve to *improve* our ability to predict and understand how the phenomena interrelate, especially in the *active* sense, and be derivatively valuable or progressive.

5. Conclusion

I have argued the following. First, it's possible for science to progress when its theories do not increase in verisimilitude. Indeed, it may progress even when they decrease in verisimilitude. Second, science does not progress centrally only, or primarily, due to increases in the verisimilitude of its theories (or accuracy of its models). Third, instead, science progresses more centrally by increasing its power to predict, and ability to furnish us with an understanding of, how the phenomena behave and interrelate. Last, but not least, power to predict concerns interventions, *inter alia*, and may be furthered, like understanding, by the development of instruments.

You need not find all of the aforementioned arguments persuasive in order to find yourself moving significantly towards cognitive instrumentalism. For instance, even if I were only successful in arguing that science progresses at least as centrally when its power to predict (or enable understanding) increases as it does when its truthlikeness increases, this coheres well with the theses I shall argue for in the next two chapters and the image of science that will emerge from the book.

This completes my initial treatment of the first element of cognitive instrumentalism. However, this element is elaborated upon and further supported later in the book. It is elaborated upon mainly in so far as the notion of understanding operating therein, which I call *empirical understanding*, is further motivated and developed (especially in Chapter 5). It is supported further by the analyses of several historical episodes (appearing mainly in Chapter 4).

Notes

1 This chapter contains material from 'Scientific Progress Without Increasing Verisimilitude: In Response to Niiniluoto', *Studies in History and Philosophy of Science* 51(1), 100–104 (2015).
2 Bird subscribes to the account of knowledge advanced by Williamson (2000).
3 Cevolani and Tambolo (2013) and Niiniluoto (2014) also criticise the view, and especially Bird's objections to their alternative.
4 As Psillos (1999: xxi) puts it: 'realists argue that an attitude towards science which involves less than aiming at theoretical truth and acceptance of the truth of theories would leave us, in some concrete respects that empiricists should recognize, *worse off* than would the recommended realist attitude.'
5 Like Lyons (2005), I take such theses not to be tied to descriptive epistemic theses about what science can reliably, and should be expected to, achieve. Even if it were to be true that science will ascertain the truth about the world at some point, for instance, it would not follow that making scientific progress consists primarily in (getting closer to) so doing.
6 I will not defend the use of thought experiments in this context, although I have reservations about their use in some circumstances. Suffice it to say three things. First, the other participants in this debate also use thought experiments (whether fully hypothetical or counterfactual in character); Bird (2007b), for example, is full of them. Second, there is no easy way to avoid using thought experiments, because even judgements about what might instead have happened in history are thought experimental (*qua* counterfactual) in character. (And some relevant historical cases, such as the discovery of the Balmer series, are covered in Chapter 4.) Third, thought experiments are important in science too, although perhaps for different reasons; see Rowbottom (2014c) for my account of these.
7 Naturally, they could remain willing to re-examine the theory in the future, in the event that new evidence, questioning its truth, should come to light.
8 See Weisberg (2007) for a discussion of the different forms that idealisations can take. See Laymon (1990) on different types of approximation, and how they relate to idealisations. There are many idealisations that I haven't listed, as comes across in the following interesting passage from Ashby (1956: 39–40):

> Our first impulse is to point at the pendulum and to say "the system is that thing there". This method, however, has a fundamental disadvantage: every material object contains no less than an infinity of variables and therefore of possible systems. The real pendulum, for instance, has not only length and position; it has also mass, temperature, electric conductivity, crystalline structure, chemical impurities, some radio-activity, velocity, reflecting power, tensile strength, a surface film of moisture, bacterial contamination, an optical absorption, elasticity, shape, specific gravity, and so on and on. Any suggestion that we should study "all" the facts is unrealistic, and actually the attempt is never made.

My thanks to Paisley Livingston for drawing this passage to my attention.
9 Diagram used according to the following licence: http://commons.wikimedia.org/wiki/User:Example (User: Krishnavedala/CC BY-SA 3.0).
10 Anderson (1997) adds that Rayleigh behaved like an engineer, rather than a physicist, in adjusting his findings. I disagree, for reasons I later explain.
11 Peters (2005: 31) also bemoans the: 'sad fact that science has not apparently progressed far beyond the earliest efforts to understand anelasticity'. But his complaint is *not* that the true equation is unknown. He refers explicitly to the failure to 'model' adequately.

12 See, for instance, von Plato (1991) on Boltzmann's ergodic hypothesis.
13 We may conclude, as a result, that simplicity and scope are truth conducive neither in isolation nor in combination when accuracy is not fixed. I here refer to three items on Kuhn's (1977: 320–339) list of theoretical virtues.
14 Friedman (1953: 14) continues, however, as follows:

> The reason is simple. A hypothesis is important if it "explains" much by little, that is, if it abstracts the common and crucial elements from the mass of complex and detailed circumstances surrounding the phenomena to be explained and permits valid prediction on the basis of them alone.

This is potentially misleading in so far as mere 'deletion' of factors is not all that occurs, and abstraction might further be distinguished from idealisation. In the words of Cartwright (1989: 354):

> In (ur-)idealization, we start with a concrete object and we mentally discount some of its inconvenient features—some of its specific properties—before we try to write down a law for it. Our paradigm is the frictionless plane. We start with a particular plane. Since we are using it to study the inertial properties of matter, we ignore the small perturbations produced by friction. Often we do not just delete factors. Instead we replace them by others which are easier to think about, usually easier to calculate with . . . most of what are called idealizations in physics are a combination of this ur-idealization with what I am calling "abstraction."
> By contrast, in . . . abstraction . . . we do not subtract or change some particular features or properties; rather mentally we take away any matter that it is made of, and all that follows from that. This means that the law we get by material abstraction functions very differently from idealized laws. . . . [W]here it is the matter that has been mentally subtracted, it makes no sense to talk about the departure of the remaining law . . . from truth, about whether this departure is small or not, or about how to calculate it. After a material abstraction, we are left with a law that is meant not literally to describe the behavior of objects in its domain, but, rather . . . to reveal the underlying principles by which they operate.

My discussion of pendulum models not involving fundamental variables, or variables concerning microscopic things, relates to this notion of abstraction, at least if it might be understood to be a matter of degree. In brief, the material *composition* of the pendulum is disregarded in the simple pendulum model.
15 As Matthews (2005: 209) concisely puts it:

> Galileo's discovery of the properties of pendulum motion depended on his adoption of the novel methodology of idealization. . . . As long as scientific claims were judged by how the world was immediately seen to behave, and as long as mathematics and physics were kept separate, then Galileo's pendulum claims could not be substantiated; the evidence was against them.

16 See also the more detailed discussion in Rowbottom (2011b, 2011c).
17 Some time after I'd written the first draft of this chapter, Dellsén (2016: 74) proposed a noetic view on which scientific progress 'consist[s] in increasing scientific understanding'. Provided an appropriate bridge principle is assumed to hold between aims and progress—such that, for instance: to achieve an aim, or get closer to achieving an aim, is to make progress—then the basis for this idea is older. For example, De Regt and Dieks (2005: 165) note that: 'Achieving understanding is a generally acknowledged aim of science.' I return to this matter in Chapter 5, where I deal with understanding in greater depth.

18 Note that the 'possible alternatives' need not be general, or also have maximal scope; for example, there could be multiple alternatives with individually less scope, but jointly as much scope. Hence, it is not necessary to assume that there is more than one theory that saves all the phenomena in the domain.

19 Niiniluoto would find this option attractive, going by the following passage:

> In my view, the axiology of science should . . . be governed by a primary rule: try to find the complete true answer to your cognitive problem, i.e. try to reach or approach this goal. Truthlikeness measures how close we come to this goal. As secondary rules, we may then require that our answer is justified, simple, consilient, etc.
>
> (Niiniluoto 2002: 174)

20 I hold that the value is purely instrumental, but consider the assumption of intrinsic value to be a generous concession (to many realists) in this context. It also streamlines the discussion.

21 I hold that achieving understanding of how phenomena interrelate is a distinct end, and not merely a means to an end. Hence, as noted in the introduction, I disagree with the following characterisation of instrumentalism due to Sober (1999: 5):

> Instrumentalism does not deny that theories are and ought to be judged by their simplicity, their ability to unify disparate phenomena, and so on. However, instrumentalism regards these considerations as relevant only in so far as they reflect on a theory's predictive accuracy. If two theories are predictively equivalent, then a difference in simplicity or unification makes no difference, as far as instrumentalism is concerned. Simplicity, unification, even fitting the data at hand are simply means to the end of securing predictive accuracy.

Even historically speaking, such a characterisation of instrumentalism is insufficiently narrow. Mach, for example, placed great emphasis on *economy* in saving the phenomena. Pojman (2009) summarises his position as follows:

> The purpose of science is to give the most economical description of nature as possible, because science is to provide conceptions which can help us better orient ourselves to our world, and if science is uneconomical then it is useless in this regard. Put another way, Mach's reason for insisting that economy must be a guiding principle in accepting or rejecting a theory is that uneconomical theories cannot fulfill their biological function, which . . . he insists is (in a descriptive sense) the function of science. The biological purpose of science is the improvement or the better adaptation of memory in service of the organism's development.

I agree with Mach that economy is important, for two reasons. First, complexity tends to make prediction difficult. Second, simplicity and understanding are linked. Both of these points are made in my case study of the pendulum.

22 This is the view defended by Strevens (2013: 510): 'scientific understanding is that state produced, and only produced, by grasping a correct explanation'. I return to this in Chapter 5.

23 The thought experiment can be adjusted so that it concerns only predictive power vs. truth, if desired.

24 One might think instead that to probe into the fundamental nature of things would be to do bad, or simply inferior, science in context. This less bold claim suffices to support the conclusion that I draw in this section.

25 As Rickles (2014: 228–229) explains:

> A rough estimate that is often suggested . . . is 10^{500} possible ground
> states. . . . This is the contemporary meaning of 'string Landscape'. . . . [and
> there are] two ways of viewing the plurality:
>
> 1. Treat the landscape's elements as corresponding to dynamical possi-
>   bilities (once necessary identifications due to dualities have eliminated
>   redundant points).
> 2. Find some mechanism or principle to break the plurality down to our
>   world.

26 Compare the following sentiments of Gell-Mann (1987: 208):

> My attitude towards pure mathematics has undergone a great change. I no
> longer regard it as merely a game with rules made up by mathematicians
> and with rewards going to those who make up the rules with the richest
> apparent consequences. Despite the fact that many mathematicians spurn
> the connection with Nature (which led me in the past to say that math-
> ematics bore the same sort of relation to science that masturbation does
> to sex), they are in fact investigating a real science of their own, with an
> elusive definition, but one that somehow concerns the rules for all possible
> systems or structures that Nature might employ. Rich and self-consistent
> structures are not so easy to come by, and that is why superstring theory,
> although not discovered by the usual inductive procedure based princi-
> pally on experimental evidence, may prove to be right anyway.

> My response is that identifying fundamental possibilities is fine—and that some
> such possibilities may prove useful for modelling purposes—but that distin-
> guishing between those possibilities with respect to truth (or truth-likeness) is
> not generally (1) reliable given our conceptual limits, or (2) required in order to
> save and understand the phenomena. See Chapter 3 on (1). I should add that
> it is remarkably optimistic to assume that we are able to determine 'the rules
> for *all* possible systems or structures', given our limitations. See Rickles (2014:
> §10.3) for more on the controversy regarding string theory.

27 For recent criticism of this view, see Stanley and Williamson (2001) and Stanley (2011). See also Bengson and Moffett (2011) and Hetherington (2011).
For a concise summary of the *prima facie* arguments, see Fantl (2012: §3.1).
For present purposes, I will steer clear of this debate. For even if *knowing
how* is a species of *knowing that*, it does not follow that the kind of *know
how* relevant in the present context concerns scientific theories, rather than
actions in the observable realm. The quotation from Baird and Faust, which
follows in the main text, illustrates this nicely.

28 I disagree with Mizrahi (2013a), however, that we should relate accounts of
scientific progress closely to the aims of scientists. As I explain in the Appendix, the aims of scientists bear no interesting relationship to what has been
misleadingly called 'the aim of science'.

29 In the words of Hacking (1983: 149):

> Philosophers of science constantly discuss theories and representation of
> reality, but say almost nothing about experiment, technology, or the use
> of knowledge to alter the world.

> Mizrahi (2013a: 384) notes that Kitcher (1993) also 'acknowledges practi-
> cal progress as a goal of science'. Devitt (2011b) uses the idea that scientific
> know how increases in an attempt to resist a central argument against the

view that contemporary scientific theories are approximately true, namely the argument from unconceived alternatives. (The basic idea is that scientists reliably get better at discovering fundamental truths.) I present and extend the argument from unconceived alternatives in Chapter 3, and respond to Devitt (2011b) in Chapter 6.

30 Another story in support is mentioned by Day (2012).

31 Furthermore, we were *convinced* that presentational hypotheses were present in at least 10% of the papers (where hypothesis testing talk appeared). See Rowbottom and Alexander (2011: 255) for the tables.

32 Naturally, the idea that problem solving is important in science is far from new. Most notably, Laudan (1977, 1996) suggests it is the main 'aim' of the activity. The notion of problem solving was also central to Popper's views on science and learning—as shown by Petersen (1984)—although this is not as widely recognised as it should be.

2   The Limits of Scientific Discourse About the Unobservable[1]

> Eddington said something to the effect that electrons were very useful conceptions but might not have any real existence. Whereupon Rutherford got up and protested indignantly, 'Not exist, not exist—why I can see the little beggars there in front of me as plainly as I can see that spoon.'
>
> —Andrade (1962: 39)

> Everything which we observe in nature imprints itself *uncomprehended* and *unanalyzed* in our percepts and ideas, which, then, in their turn, mimic the processes of nature in their most general and most striking features. In these accumulated experiences we possess a treasure-store which is ever close at hand and of which only the smallest portion is embodied in clear articulate thought.
>
> —Mach (1893: 36)

In the previous chapter, I argued that science is valuable primarily in so far as it enables us to save (a significant proper subset of) the phenomena, and furnishes us with an understanding of how those phenomena interrelate. Finding true or approximately true theories would be a nice accompaniment, but is not a part of the main dish.

In the present chapter, I will turn my attention to semantic concerns, and particularly to how discourse concerning unobservable things—about physical things 'beyond' the phenomena—fits into the picture. First and foremost, I will defend a position that I call *property instrumentalism*. The key idea involved is that talk of unobservable objects should be taken literally *only* in so far as those objects are assigned properties, or described in terms of analogies involving other things, with which we are experientially (or otherwise) acquainted. Second, I will argue that *intentionally* non-literal discourse about unobservable things is important in science.

1. Semantic Instrumentalism: A Moderate Version

Traditional instrumentalism of a semantic variety involves the claim that *all* scientific discourse about unobservable things should be taken non-literally.[2] ('Literal' is here opposed to 'metaphorical', not to 'false'.) This view is now extremely unpopular, and is rarely discussed in any serious depth, except when the object of the exercise is to show why it is refuted.[3] This is part of a widespread backlash against the linguistic turn in philosophy. Van Fraassen (1980: 56) even goes so far as to declare:

> The main lesson of twentieth-century philosophy of science may well be this: no concept which is essentially language-dependent has any philosophical importance at all.

This conclusion is too bold, however, because the core idea behind linguistically motivated instrumentalism, such as that of the logical positivists and Ayer, remains plausible. How can we name something with which we are not acquainted—where to be acquainted with an entity is to 'have a direct cognitive relation to' or be 'directly aware of' the entity (Russell 1911: 108)—unless that name is a substitute for a definite description made in terms of names for entities with which we *are* acquainted (or other names reducible to such descriptions)? It appears we cannot. In the words of Searle (1958: 168):

> How . . . do we learn and teach the use of proper names? This seems quite simple—we identify the object, and assuming that our student understands the general conventions governing proper names, we explain that this word is the name of that object. But unless our student already knows another proper name of the object, we can only *identify* the object . . . by ostension or description.

So even if one disagrees with Russell (1918) that proper nouns should be understood as *substitutes* for descriptions, using such words in a referential capacity—to 'denote the individuals who are called by them' (Mill 1843: bk.I, ch. 2, §5)—requires appropriate underpinning acquaintances. (This could be an acquaintance with the referent, or with the referents of terms used in a description—not necessarily a definite description—picking out the referent in some context.)

  Moreover, theoretical terms are typically kind (or type) terms such as 'electron', and there is a description (e.g. involving a specification of mass, charge, and spin) with which such terms are associated. So how could we understand such terms completely if we were not acquainted with the referents of the property names mentioned in the associated description? Surely we could not. That is, although an incomplete understanding may be possible in the presence of acquaintance with only some of the referents of the property names. For example, someone born deaf might grasp the structure of Tchaikovsky's first symphony, including the durations of the notes, without grasping any of the ways it might sound when played correctly. (I say 'ways' because the sounds generated by different instruments and players vary, while remaining consistent with the score.) And the beauty of the symphony might be appreciated to some extent by such a deaf aficionado, by the use of appropriate analogical thinking.

But accepting a fundamental link between naming and acquaintance does not entail accepting semantic instrumentalism. First, one must add the empiricist thesis that our primary means of becoming acquainted with things—both objects and the properties thereof—is sensory experience. Second, one must add that what we are acquainted with is limited in such a way as to render it impossible to comprehend or discuss (some significant aspects of) any truth behind the appearances.

The first thesis is reasonably uncontroversial; most of us are content with the idea that sensory experience is a major source of acquaintance, that there are many things (such as colour and pitch) with which we would never become acquainted in the absence of sensory experience, and that the unobservable entities discussed by science are normally taken to have observable effects (by which we are supposed to infer their existence). Note also that the semantic instrumentalist need not rule out the view that we are acquainted with some things by means other than direct personal experience, e.g. via innate phylogenetic (or even ontogenetic) knowledge. Against Mach (1893), it is even possible to accept intuition as an experience-independent source of acquaintance while defending the view that we should not, in general, construe scientific theories literally. It is incorrect that acceptance of the thesis that we can discuss only that with which we are acquainted 'leads inevitably into phenomenalism' (Maxwell 1962: 12).[4]

I shall therefore focus on the second thesis in what follows, and argue for a moderate form of semantic instrumentalism. It involves a denial of semantic realism in so far as this pertains to talk of unobservable *properties*, but not unobservable objects provided that these are defined in terms of observable properties or by analogy with observables. I will flesh out this position, which I call *property instrumentalism*, subsequently. But to avoid any early confusion, it should be emphasised that an entity may possess an observable property without the property being observable *in that particular instance*. The language of metaphysics, which I use purely for illustrative purposes, helps here. The trope of red of a rose in my garden may be observable, although the trope of red of a tiny insect may not be. Yet the unobservable insect may nevertheless instantiate an observable property, namely redness, in so far as some of its instances may be observed. We may see red roses, for example.

Now to see how a position like property instrumentalism is typically unanticipated, consider part of Psillos's (1999: xix) definition of the semantic thesis of scientific realism:

> Theoretical assertions are not reducible to claims about the behaviour of observables, nor are they merely instrumental devices for establishing connections between observables. The theoretical terms featuring in theories have putative factual reference.

Agreement with the first sentence does not signal agreement with the second. Rather, as we will see, it is possible for an instrumentalist to accept that some theoretical terms 'have putative factual reference' but deny that all have it. Moreover, it is possible for an instrumentalist to defend the view that some 'theoretical assertions' are instrumental devices—to promote understanding of relationships between phenomena, in line with the notion of progress defended in the previous chapter, *inter alia*—and even that such devices are indispensable for good science.

Psillos (1999) is not peculiar in this oversight. The semantic instrumentalist may equally deny what Sankey (2008: 14) instead calls 'theoretical discourse realism', namely that 'scientific discourse about theoretical entities is to be interpreted in a literal fashion as discourse which is genuinely committed to the existence of real unobservable entities', without denying that *some* discourse about theoretical entities should be interpreted literally. As such, it is incorrect to say that 'Instrumentalism denies the literal interpretation of theoretical discourse' (ibid.); rather, it involves the denial of the literal interpretation of *some* (significant part of) theoretical discourse. If one wishes instead to insist on Sankey's definition of instrumentalism, then one must admit that it is possible to deny 'theoretical discourse realism' (or semantic realism) without being an instrumentalist (or a reductive empiricist), and without denying the metaphysical thesis that, in the words of Psillos (1999: xix), 'the world has a definite and mind-independent natural-kind structure' or a correspondence theory of truth.[5] Either way, there is sparsely inhabited intellectual territory where sweeping generalisations about theoretical discourse are rejected in place of finer-grained distinctions. The property instrumentalist dwells in this territory, and I call the position 'instrumentalist' because it shares a key feature of the (more radical) positions that have historically borne the name. Specifically, it retains the idea that the distinction between observable and unobservable *has significance with respect to meaning*. It is novel in so far as it focuses largely on properties, rather than property bearers.

Admittedly, what's sauce for the goose is sauce for the gander. Semantic realists need not endorse the extreme positions mentioned previously—i.e., Psillos's 'semantic thesis' or Sankey's 'theoretical discourse realism'—which I use as foils for property instrumentalism. It would be rather extreme to deny that *some* theoretical discourse in science appears only as a matter of convention, or for pragmatic reasons (even when scientists fail to make this explicit or to realise this). For example, even the most ardent realist should accept that electrons don't really have *negative* charges (as opposed to charges that are *different in kind* from those of some other entities, like protons). For the point behind using 'positive' and 'negative' designations is mathematical convenience. Multiplying two numbers with the same sign results in a positive, whereas multiplying two numbers with different signs results in a negative, and this is helpful for characterising force directions (as 'repulsive' or 'attractive'). Hence, replacing all positives with negatives, and vice versa, leads to exactly the same predictions without altering the number of posited property types or tokens.

A less ardent realist might also accept that there's no literal sense to the idea of another possible world where the charges on all things classified as 'negative' here and all things classified as 'positive' here are swapped, but everything else remains the same, such that the initial conditions of the universe and its evolution might be specified in precisely the same way, in the same natural and mathematical language, as they are here. Consider dispositional essentialism, or the view that all properties are dispositions. If one is metaphysically committed to this—as some scientific realists, such as Bird (2007a), are—then one must take the two possible worlds to be the same. That's because charge is specified in terms of dispositions to attract and repel rather than some categorical property that can be swapped around. Any electron has the disposition to repel other electrons, attract protons and positrons, and so forth.

A semantic realist of an intermediate variety might buy into charges being intrinsic and different, but concede that centres of mass and instantaneous velocities do not exist (although terms corresponding to these appear in scientific equations). Butterfield (2014), for instance, takes the possibility of a world where charges are 'swapped', as mentioned previously, ontologically seriously. But he inveighs against pointillism—'the doctrine that a physical theory's fundamental quantities are defined at points of space or of spacetime' (Butterfield 2006)—in classical mechanics.

So semantic realists have considerable room for movement. Nonetheless, I take semantic realists to think that talk of electrons and quarks, and other unobservable entities, should be taken literally to an extent that I deny. In short, that's to say, I take there to be a significant difference between property instrumentalism and any version of semantic realism that's watered-down enough from the extreme versions to be a promising alternative.

Note also that a property instrumentalist need not agree with the view on scientific progress outlined in the previous chapter, although as I'll show in Chapter 4, with examples from the history of science, property instrumentalism coheres with such a view on progress. To adopt property instrumentalism is not to adopt any particular view on scientific progress (or the value of science), although it *is* to rule out the view that (the whole) truth can be found if there are unobservable things possessing properties of kinds with which we are not, and cannot become, acquainted.

2. Articulating Property Instrumentalism

But why should we think that unobservable things are different from observable things, with respect to the kinds of properties that they bear? Perhaps one might make the case that some of the difficulties we've encountered with the interpretation of quantum mechanics support this view. For instance, one might agree with Bohr that we are bound to use classical concepts that we cannot apply to quantum objects directly, but may instead employ only as complementary descriptions on a context-by-context basis (Faye 1991, 2008).[6] Similarly, Ladyman (1998: 422) suggests that 'The demand for an individuals-based ontology may be criticised on the grounds that it is the demand that the structure of the mind-independent world be imaginable in terms of the categories of the world of experience.'[7] Yet even if we instead admit that there is no evidential reason to expect unobservable things to differ from their observable counterparts, there appears equally to be no evidential reason to expect them all to be similar.

There are also old worries about how the observable should be distinguished from the unobservable, and whether the latter can become the former, but suspend these for the time being. We will return to them in Chapter 6, where the observable-unobservable distinction—which underpins many theses in this book—is discussed at length. For the moment, suffice it to say that the property instrumentalist may admit that the distinction is vague, yet emphasise that there are clear examples on each side (such as rabbits and quarks).[8] And she may say that we should look to science itself to work out which properties are observable (or observed), and therefore to judge whether to take a given theory literally (van Fraassen 1980: 59).[9] Whether some particular theoretical discourse may be taken literally is therefore an empirical question.

I will now turn my attention to explaining the property instrumentalist's stance on theoretical discourse in a little more depth, with the aid of some historical examples. There are two principal ways in which one may posit unobservable things by appeal to observables: by employing observable properties, or by employing analogies with observable things. I will discuss each in turn.

2.1 Employing Observable Properties

The first way to posit unobservable things by appeal to observables is to use observable properties. Sometimes this may be done by straightforward extrapolation. When I hold an apple in my hand, I feel the force that it exerts. But if I halve the apple, and so on, I will eventually reach a point at which there is no discernible force. It does not appear unreasonable, however, to presume that there is a force acting, and that said force is in proportion to the amount of apple remaining (as is the case when the forces are discernible).[10] There is an upper limit to the force I can discern too. The difference between having my hand crushed by a car and crushed by a lorry is not directly sensible. We must therefore extrapolate 'upward' into some parts of the observable realm as well.

Although he fails to recognise that 'upwards' extrapolation may also be required, Mach (1893: 588) provides similar examples of the 'downward' process, concerning the frequency of sound and the amplitude of vibrations:

> Even when the sound has reached so high a pitch and the vibrations have become so small that the previous means of observation are not of avail, we still *advantageously* imagine the sounding rod to perform vibrations. . . . [T]his is exactly what we do when we imagine a moving body which has just disappeared behind a pillar, or a comet at the moment invisible, as continuing its motion and retaining its previously observed properties. . . . We fill out the gaps in experience by the ideas that experience suggests.

The property instrumentalist differs from Mach in taking such descriptions literally. Unlike Mach, she is not committed to the view that 'the world is not composed of "things" as its elements, but of colors, tones, pressures, spaces, times, in short what we ordinarily call individual sensations' (Mach 1893: 579) or that '[w]hat we represent to ourselves behind the appearances exists *only* in our understanding, and has for us only the value of a *memoria technica* or formula' (Mach 1911: 49). Nor is she committed to what Brush (1968: 197) appropriately calls 'empiriocriticism', namely 'a critical view toward all scientific hypotheses not directly induced from experiment.'

The history of atomism provides many examples of the ascription of observable properties to unobservable things.[11] For example, the atmospheric atomic model of the eighteenth and nineteenth centuries might be taken literally in so far as it involves two different kinds of atoms, matter-atoms and ether-atoms, attracting one another and repelling one another (via inverse square laws) respectively. And similarly, the kinetic theory of gases might be understood literally in so far as it did not 'rely directly on a detailed atomic theory', and required 'little more information about atoms than their sizes' (Brush 1968: 195). According to property instrumentalism, extrapolation of sizes and attraction/repulsion by inverse square laws—demonstrable in electrostatic contexts involving charged macroscopic objects, for example—need not be construed as an artifice. The inverse square character of several fundamental force laws accords well with experience, and the key to understanding why is that the area of a sphere is a function of the square of its radius. Consider an analogy between an isotropic source and a charge (or mass). If the power of the source is P, then the intensity of whatever it emits at any other point is $P/4\pi r^2$, where r is the distance between that point and the source.[12] Thinking in terms of flux and concentric spheres illustrates why.

The previous examples involve property types that were observed before being ascribed to (hypothetical) unobservable entities. However, it is also possible to ascribe *unobserved yet observable* property types to unobservable entities. And property instrumentalism allows for the legitimacy of so doing. To see why this may be appropriate, consider dispositional properties such as: ductile, sonorous, malleable, fragile, and (electrically) conductive. A simple way to analyse these—simple enough for present purposes, and defended by Choi (2006, 2008) in the face of several interesting putative counterexamples[13]—is as follows:

> An object O is disposed to M when C if and only if it would M if it were the case that C.

Now imagine O is unobservable, and C is observed (and hence observable). M may be observable yet *unobserved*. Consider a concrete example. Let O be a tiny sphere. Let C be 'it collides with anything'. And let M be 'split into two spheres, each of half the volume of the original'. Then we may speak of our tiny sphere being disposed to split into two tinier spheres, each of half the volume of the original, when it collides with anything. We may take that talk literally, despite the fact that we have never encountered a manifestation like M. It's evident that it's an observable property. We can (clearly and distinctly) visualise something exhibiting it. Similarly, it's evident that unicorns, centaurs, and hippogriffs are observable but unobserved *entities*.

I do not intend to argue that any dispositional properties are observable, or even that counterfactual conditionals are truth apt. This is partly because of my empiricist leanings. As Choi and Fara (2012) put it:

> nothing about the actual behavior of an object is ever necessary for it to have the dispositions it has. Many objects differ from one another with respect to their dispositions in virtue of their merely possible behaviors, and this is a mysterious way for objects to differ.

And as Hájek (Manuscript) puts it, counterfactual conditionals 'involve a modality that makes empiricists uncomfortable'. That's why Carnap (1936/7) endeavoured to account for dispositional terms by using so-called reduction sentences, which involve indicative conditionals, instead. An example of such a sentence is: 'if any thing x is put into water at any time t, then, if x is soluble in water, x dissolves at the time t, and if x is not soluble in water, it does not' (Carnap 1936/7: 440–441).

So suffice it to say that property instrumentalism is consistent with some dispositional properties being observable. But for all it says, dispositions may fail to exist. Similarly, property instrumentalism is consistent with a wide variety of different views on the laws of nature. It's compatible with laws involving non-logical necessity, e.g. due to connections obtaining between universals (following Dretske [1977], Tooley [1977], and Armstrong [1983]). But it's also compatible with laws being exceptionless regularities with significant non-modal properties (following Ayer [1956]). It is intentionally silent on such esoteric metaphysical matters.[14]

2.2 Employing Analogies

The second way to posit unobservable things by appeal to observables is to employ analogies. Such analogies are common across all the sciences. A well-known historical example in physics is the comparison between the atom and Saturn made by Nagaoka (1904). Less well-known analogies were used in developing and critiquing earlier atomic theories. For example, Kelvin (1867) likened atoms to smoke rings.[15] Similar analogies remain commonplace in contemporary physics. For instance, Hawkins and McLeish (2004) model repressor protein dimers as rigid plates connected by springs, in order to predict how they move. See Figure 2.1.

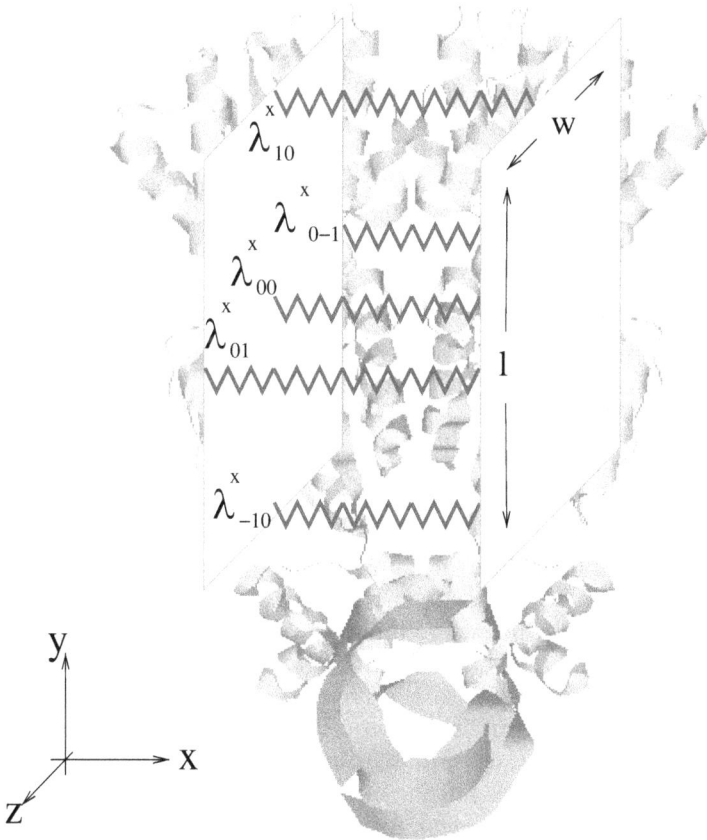

Figure 2.1 Plates and Springs Model of a Dimer

Reprinted figure with permission from Hawkins, R. J. and T. C. B. McLeish, *Physical Review Letters* 93, 098104 (2003). Copyright (2003) by the American Physical Society.

The significance of analogies is also illustrated by the presence of concrete models, in addition to abstract (or theoretical, non-material) models, in science. Saturn was just such a model, although it would have been a better model if we were able to tinker with it, to manipulate it, in order to see the effects. (Smoke rings were superior in this respect. Kelvin [1867] was interested in smoke ring interactions determined by experiment.) This is why many concrete models are artefactual, rather than natural. Artefacts can be deconstructed, reconstructed, duplicated, refined, and even, in some cases, animated.

Many more examples of concrete models appear later in this chapter. But presently we are concerned with how they interact with abstract models. For a flavour of this, consider a simple scenario. Imagine that you want to understand how some polymer—polyethylene, for instance—moves around at the molecular level. You try to visualise the situation, and you find yourself thinking of little chains. Each monomer is like a link in a chain, you suppose. Each is constructed of the same material, is roughly the same size and shape, and is connected to the others in the same way. On the basis of these considerations, you then construct an abstract model of how the polymer moves at the molecular level.

So how do concrete models fit in? You're likely to be familiar with the behaviour of a variety of chains in mundane scenarios—on necklaces, bicycles, anchors, gates, and so forth. And you might make conscious recourse to past experience of such objects in generating your abstract models. Moreover, the modelling process may rely on such concrete models even in the absence of such conscious recourse. First, we might make *unconscious* recourse to them in the initial generative process. I take this to be part of what Mach (1893: 36) claims, in the quotation with which this chapter begins: 'our . . . ideas . . . mimic the processes of nature in their most general and most striking features. . . . [W]e possess a treasure-store which is ever close at hand and of which only the smallest portion is embodied in clear articulate thought.' That's to say, the initial model of the chain may necessarily be based on our experience of chains. This is a speculative thought, especially in so far as it involves unobserved, and presumably unobservable, mental processes. But it fits with what many psychologists say about mental models, as we will see later. As a result, some realists will be inclined to think we have considerable evidence for the claim.

Second, we might make conscious recourse to familiar concrete models at a later stage in the modelling process. Think about how an abstract model might fail to be sufficient to its intended task, and what you might do if it proved to be so. On the one hand, your abstract model might fail to properly represent—and/or capture—the behaviour of observable chains. It might fail, for example, to effectively explain and/or predict the behaviour of the watch chain in your pocket as you jog (when conjoined with available auxiliary hypotheses). In this scenario, the observable chains might nevertheless highly resemble the polymer chains (in relevant respects): the analogy could be 'fit for purpose', with respect to your goal in generating the model.[16] So the abstract model would require refinement. And as part of the refinement process, you might explore the behaviour of several actual chains in a number of experiments. On the other hand, the abstract model might explain and predict the behaviour of everyday chains, but fail to be applicable to the polymer (in the desired respects). In this event, the correct way to proceed would be to alter the model to take into account relevant differences between the polymer and everyday chains. And as part of this process, you might want to construct different types of chain—and/or take pre-existent chains—and tinker with them in various ways in order to examine the effects. You might also want to consider whether your model better describes other polymers, and consider how those resemble the polymer you were first interested in.

There are also distinct cases in which a concrete model is used to animate an abstract model. Phillips' hydraulic model of the economy, where water flow is analogous to cash flow, is a prime example. Phillips found it difficult to grasp the relationships between the variables in the mathematical models he was using, and hence difficult to determine their economic consequences. Thus he cobbled together a mechanical system—around three cubic metres in size, and made of a variety of scavenged parts, including perspex, fishing line, and a windscreen wiper motor from a Lancaster bomber—in which these mathematical relations were instantiated. One story involving this model is especially amusing. Phillips reportedly asked one of his students to control interest rates (as the Bank of England does), and another of his students to control taxes and spending (as the Chancellor of the Exchequer does). The result was a wet floor.[17]

Mundane analogies are even the explicit subject matter of papers. Alexander (1999), for example, compares chemical plants to digestive systems: 'Our jaws are grinding machines, our guts are chains of biochemical reactors, our capacities to digest food and absorb the products are designed with substantial safety margins, and it may be useful to think of a mouse as a pilot plant for an elephant'. It would also be extremely difficult to teach science without recourse to analogies concerning familiar macroscopic entities and systems. Hence, many readers will remember having electric flow in a wire compared to water flow in a pipe or car flow on a road, with resistance being compared to pipe diameter or number of lanes, and so on. Thus, in summary, there is considerable support for the conclusion, drawn by the psychologist Dunbar (2002: 159) on the basis of his observations of scientists, that '[M]undane analogies are the workhorse of the scientific mind'.

The idea is not new. Smith (1980: 107), writing in the late eighteenth century, emphasised the importance of analogies—and indeed, the ultimate significance of acquaintance—in fostering understanding of the form I introduced in the previous chapter:

> To introduce order and coherence into the mind's conception of this seeming chaos of dissimilar and disjointed appearances, it was necessary to deduce all their qualities, operations, and laws of succession, from those of some particular things, with which it was perfectly acquainted and familiar, and along which its imagination could glide smoothly and easily, and without interruption. . . . [I]f any analogy could be observed betwixt the operation and laws of succession of the compound, and those of the simple objects, the movement of the fancy, in tracing their progress, became quite smooth, and natural, and easy.[18]

Other eminent scientists—especially those working in what I'll later describe as the Victorian tradition—have said similar things since. Fitzgerald (1888: 168–169), for example, wrote:

> All reasoning about nature is . . . in part necessarily reasoning from analogy. . . . Notwithstanding the danger of our mistaking analogies for likenesses there is a great advantage in studying analogies. . . . [I]f the forms of energy were as familiar a conception as eggs and money, people would find it as easy to reason about its [sic] transformation as they are [sic] about the number of eggs the old woman brought to the market and sold at one dozen at 3 a penny and so forth. It is because at every turn people lose the thread of the argument by the difficulty of realising what they are dealing with that they find it difficult at first to reason consistently about a new subject. It is on account of this that it is worth while studying analogies between things with which we are not familiar and those with which we are.[19]

In the psychological literature on mental models and analogies—see, for instance, Gentner (1983) and Blanchette and Dunbar (2000)—'superficial' similarity is distinguished from structural similarity. The former involves similarity in intrinsic properties, such as mass and absolute volume, whereas the latter, which Gentner (1983) argues is the basis of the most important analogies used in science, involves similarity in internal relations. When we compare an atom to the solar system, for example, we suggest that the nucleus corresponds to the sun, that the electrons correspond to planets, and that gravitational force corresponds to electromagnetic force. But we do not thereby wish to suggest that the nucleus (necessarily) emits radiation, that electrons have a range of masses and compositions, and so forth. What we wish to convey is that the nucleus *is much bigger than* the electron, that the electrons *orbit* the nucleus because they *are attracted to* the nucleus, and so on.

It is too strong to label the former kind of similarity as 'superficial', however, because it plays a significant role in explaining the success of some models. Consider, for instance, a billiard ball model of gases. Here, the molecules may profitably be taken to share several intrinsic properties with the balls. These include volume, translational degrees of freedom, and rotational degrees of freedom. Granted, one might think that degrees of freedom aren't intrinsic properties, in so far as degrees of freedom may be constrained in appropriate circumstances. But one speaks of 'free particles' in physics, and might equally speak of 'free billiard balls'. The intrinsic properties of such billiard balls are highly significant in models where the only interactions between the billiard balls are collisions, for example.[20]

Instead of focusing on the distinction between 'structural' and 'superficial' analogies, it is therefore preferable to consider Hesse's (1966) earlier distinction between positive, negative, and neutral analogical aspects of models. She explains this distinction as follows, with reference to the same example:

> [B]illiard balls are red or white, and hard and shiny, and we are not intending to suggest that gas molecules have these properties . . . the relation of analogy means that there are some properties of billiard balls which are not found in molecules. Let us call those properties we know belong to billiard balls and not to molecules the *negative analogy* of the model. Motion and impact, on the other hand, are just the properties of billiard balls that we do want to ascribe to molecules in our model, and these we can call the *positive analogy*. Now the important thing about this kind of model-thinking in science is that there will generally be some properties of the model about which we do not yet know whether they are positive or negative analogies; these are the interesting properties, because . . . they allow us to make new predictions. Let us call this third set of properties the *neutral analogy*.
>
> (Hesse 1966: 8)

This passage is imprecise in two significant respects, which deserve comment to dispel any resultant confusion. First, Hesse discusses in her definitions, variously, what we are 'intending to suggest', what we 'want to ascribe', and what we 'know'. Compare the definitions for 'negative analogy' and 'positive analogy'. The former involves those 'properties we *know* belong to billiard balls and not to molecules', whereas the latter involves 'properties of billiard balls that *we do want to ascribe* to molecules'. (You will also note that 'we are not intending to suggest' appears in the way that the former is set up, which might follow from, but need not be a result of, knowing that the relevant properties are not shared.) Now one option, *prima facie*, would be to cast all the talk in terms of knowledge. But this won't do. Consider just the positive analogy, in order to see this. Scientists often generate models that they don't know, or even strongly believe, will be adequate for their intended purposes. So the positive analogy, in such cases, would be better said to involve the properties that the scientists 'want to ascribe' to the target, or are 'intending to suggest' are present in the target. But this is not quite right either. The reason is that scientists sometimes consider property ascriptions to be *functional* in making a model serve its intended purpose *without taking those property ascriptions to be true*. One uncontroversial way in which this happens—on which realists and anti-realists may agree—is when the target is considered only to *approximately* instantiate the property present in the model. It might also happen—somewhat more controversially—when scientists anticipate that thinking in terms of the presence of a property will be productive, irrespective of whether it's present. As a result, the *positive analogy* should be recast as involving the set of properties that the modeller expects to do some work (in furthering predictive power, understanding, or what have you). The other two kinds of analogy should be recast along similar lines, as follows. The *negative analogy* involves the set of the properties that the modeller expects not to do any such work.[21] Finally, the *neutral analogy* involves the remainder of the properties, about which the modeller is unsure.

This brings us to the second imprecision, which I have already corrected. It occurs when the neutral analogy is said to involve 'properties of the model about which we do not know whether they are positive or negative analogies'. But clearly, Hesse did not intend to suggest that we do not know what we intend (or know what we know).

Hesse used 'know' on so many occasions because she presumed some form of scientific realism. But the way I have described 'the analogies' is less presumptuous. Consider, again, positive analogies 'as involving the set of properties that the modeller expects to do some work'. These may be fruitful even when said properties aren't possessed, or even approximately possessed, by the target. For instance, they may give us a neat way to think of the target, or to predict the observable effects of the target in a variety of situations, in the framework of the theories underlying the model. More veridical models may be less useful in the aforementioned respects, say if the theories underlying the model are false. In short, the situation here is analogous to that in deductive logic. Although true premises guarantee true conclusions, untrue premises do not guarantee false conclusions.

Often, moreover, many different analogies are used to describe the same thing *in different contexts*. We have already touched on Bohr's notion of complementarity, and the idea that we can conveniently think of quantum entities as waves at some points (e.g. when they move), and particles at others (e.g. when they exchange energy), without thinking of them as truly having 'wave-particle' natures. In this regard, it is crucial to note the limited scope of appropriate analogical statements. 'Electrons are like waves' is erroneous. 'Electrons move like waves' may not be.

Analogies may be taken literally in so far as such derivative analogical statements have truth conditions. They are true when appropriate *resemblances*, structural or otherwise, are in place. For example, it may be true that the sodium and chlorine ions in solid common salt are arranged similarly to how cannonballs were once typically stacked—after Harriot's investigation, at Raleigh's request, of the most efficient way to do so—and oranges are still stacked by greengrocers (i.e., in a face-centred cubic manner).[22] However, it is evident that Cassius Clay did not float like a butterfly or sting like a bee.

To say that such analogies *may* be taken literally to some extent isn't to say that they always should be, however. This brings me on to the next section.

2.3 Chemical Models, 1861–1874: A Case Study on Property Instrumentalism and Localism

> Dalton used square wooden blocks of different colours for illustrating his atomic theory, and it happened indeed, that a dunce, when asked to explain the atomic theory, said: "Atoms are square blocks of wood invented by Dr. Dalton."
>
> —Schorlemmer (1894: 117)

Near the start of this chapter, I urged that 'sweeping generalisations about theoretical discourse [be] rejected in place of finer-grained distinctions'. And I should like to emphasise, at this juncture, that property instrumentalism involves no pernicious generalisations in so far as *it posits only necessary, and not sufficient, conditions for scientific discourse to be construed literally*. Thus it fits with Gardner's (1979: 32) conclusion that:

> Those who maintain that all scientific theories are devices for predicting observable events, and those who say all scientific theories purport to be literally true, are both mistaken. Moreover, which is true of a theory may change with time.

In defending this conclusion, which I've already independently motivated, Gardner (1979) follows in the footsteps of Shapere (1969, 1974). He writes:

> Shapere [(1974: 120)] notes that a common pattern in the history of science is that a theory is first put forward or received as merely an idealization or calculational device, and then later comes to be regarded as literally true. His examples are Galileo's theory of the

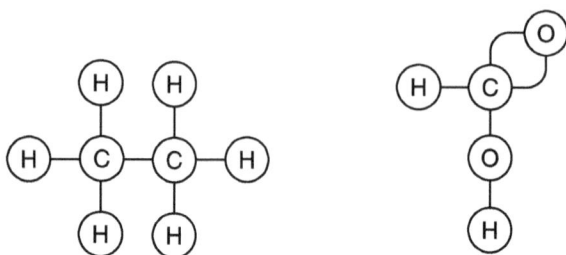

Figure 2.2 Crum Brown's (1864) Graphical Formulae for Ethane and Methanoic Acid

parabolic path of projectiles . . . and Planck's theory of the quantum of energy.

(Gardner 1979: 3)

Gardner (1979) uses the history of atomism to add grist to Shapere's mill. But since I will discuss this in Chapter 4, I will here provide a different example of the 'common pattern' and show how it fits with property instrumentalism. This comes from the history of chemistry, and centrally concerns concrete models.

In 1864, Crum Brown presented a new graphical way to represent the composition of molecules.[23] Figure 2.2 shows how he depicted two simple molecules.

This model was purely abstract, or theoretical, in character. Like several preceding models—on which see Rocke (1984)—it was 'regarded as a mere aid to classifying reactivities and searching for analogies—a taxonomic model with no correlation to a reality that was assumed to be fundamentally unintelligible' (Meinel 2004: 244).[24] In the words of Ritter (2001: 39): 'ontological skepticism . . . was at the heart of chemical atomism.'[25]

Just a year later, however, Hofmann (1865) devised a concrete correlate to graphical notation, which he employed in a lecture to an aristocratic audience:

> Croquet balls were Hofmann's atoms, painted in white for hydrogen, green for chlorine, red for the 'fiery oxygen', blue for nitrogen, and black for carbon—colour codes still in use today which seem to originate from that remarkable evening lecture.
>
> (Meinel 2004: 250)

The balls could be connected by thin metal bars, and each coloured ball had an appropriate number of holes, or connection points, into which to insert these: black balls had four, blue balls had three, and so on down, in line with the maximum number of bonds we now think of

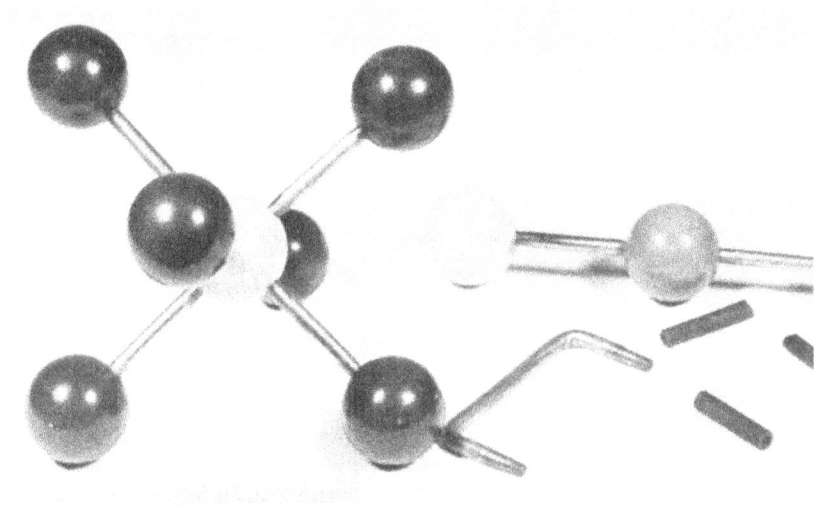

Figure 2.3 'Glyptic Formulae' Concrete Model Set[26]

Image reproduced with permission directly from Hill, C. R. 1971. *Chemical Apparatus.* Oxford: Museum of the History of Science. Inv. 42347. © Museum of the History of Science.

the corresponding types of atoms forming. There were two types of thin metal bars, one curved and one straight. Figure 2.3 shows part of a set of the same style, which was marketed shortly thereafter.

The added spatial dimension in Hofmann's model was merely a result of the medium of representation. As Meinel (2004: 252) puts it: 'its spatial properties were clearly not a consequence of theoretical considerations, but a mere side effect of using croquet balls to turn lines on paper or a blackboard into a mechanical device to be put on the table of the lecture theatre'. In other words, the spatial properties of the model were part of the *negative analogy*; Hofmann did not expect these to do any useful scientific work.[27] It should therefore be of little surprise that these concrete models, of which few remain, appear to have been used exclusively for teaching purposes. It is also notable that some scientists objected to their use precisely because they feared that students would not understand that the spatial properties were part of the negative analogy. Interestingly, Crum Brown (1864) even anticipated this worry concerning his graphical notation.

However, it did not take long before such concrete models, and hence their theoretical counterparts, began to be seen rather differently. The story of how this happened is rather complex, and involves several independent developments. Here are just three parts of the story, which I present in no particular order. First, in 1867, Kekulé, Körner and Schubart developed a concrete model that changed the orientation of the connection points on the 'atoms', and therefore allowed new model geometries. Black balls were now connected with others in a *tetrahedral* manner, whereas they had previously been connected in a square planar manner, for instance.

Second, Paternò (1869) drew a conclusion about carbon-carbon double bonds by analogy with the aforementioned concrete model, namely that these restrict rotation (in a way that single bonds do not). On this basis, he predicted that halogenated ethanes have two isomers; the halogen atoms may either face in the same direction or in opposed directions, with respect to the double bond.

Third, Frankland (1870) developed a subtly new means of graphical notation—where double bonds were represented by twin straight lines (and the circles around the letters were removed)—and started to refer to the connections as 'bonds'. Frankland (1870: v) stated that his 'system of symbolic notation . . . is so framed as to express the same ideas, of the chemical functions of atomic, as the graphic [i.e., Crum Brown's] and glyptic [i.e., Hofmann's] formulae, with which . . . it harmonizes completely'. However, he took some elements of such models to be part of a *positive* analogy:

> GRAPHIC NOTATION . . . is invaluable for clearly showing the arrangement of the individual atoms of a chemical compound. . . . [T]he graphic form affords the most important assistance, both in fixing upon the mind the true meaning of symbolic formulae, and also in making comparatively easy of comprehension the probable internal arrangement of the very complex molecules frequently met with both in mineral and organic compounds. It is also of especial value in rendering strikingly evident the causes of isomerism in organic bodies. . . .
>
> [T]hese graphic formulae are intended to represent neither the shape of the molecules, nor the supposed relative position of the constituent hypothetical atoms. The lines . . . serve only to show the definite disposal of the bonds . . . thus the formula for nitric acid *indicates that two of the three constituent atoms of oxygen are combined with nitrogen alone*. . . .
>
> The lines connecting the different atoms . . . are but crude symbols of the bond of union between them; and it is scarcely necessary to remark that no such material connexions exist, the bonds which actually hold together the constituents of a compound being, as regards their nature, entirely unknown. [emphasis added]
>
> (Frankland 1870: 23, 25)

So Frankland took the bonds to be real in some sense, even if he took them to be immaterial. The rods in the concrete models, like the lines in their abstract counterparts, represented *something*.[28] Moreover, it is plausible that some student readers of Frankland (1870) would have taken the spatial orientation to be a *neutral*, rather than *negative*, part of the analogy. That the formulae were not 'intended to represent . . . the shape of the molecules' is consistent with them representing the shape nonetheless (although the subsequent use of 'serve only' suggests that Frankland took the spatial orientation to be part of the negative analogy).

To explore exactly how these developments interrelated would take us too far astray. (And my story is, alas, incomplete. For example, I have not covered the work of Wislencius.) However, it is plausible that each contributed somewhat towards taking the spatial aspects of molecular models literally. Overall, the change happened quickly. In 1866, Kolbe wrote to Frankland that 'graphic representations are . . . dangerous because they leave too much scope for the imagination. . . . It is impossible, and will ever remain so, to arrive at a notion of the spatial arrangement of atoms' (Rocke 1993: 314). In 1874, Le Bel and van't Hoff—the latter worked with Kekulé in 1872 and 1873—both presented 'stereochemical theories were intrinsically spatial, because their explanatory power depended precisely on their describing the arrangement of atoms in space' (Hendry 2012: 297). So '[s]tereochemists . . . believed they had access to the physical appearance of the molecule, and had not simply invented instruments for prediction' (Ramberg 2003: 328).

Some authors think the change over this short period was due mainly, or exclusively, to the introduction of the concrete models:

> New views do not emerge at once, nor do they spring from a single discovery. The chemists prepared 3-D models for teaching purposes, and in using them they learned to link the mind's eye with theoretical notions, with the manipulating hand, and with laboratory practice. . . . The introduction of 3-D molecular models was not exclusively, nor even predominantly, part of a theoretical discourse, as it often assumed in the literature. Instead, the models were primarily used as tools for the creation of new types of *Anschauung* not only in the audiences taught, but also in the minds of those who developed these tools in struggling with the growing complexity of chemical constitution.
>
> (Meinel 2004: 265–266)

However, an appropriate conceptualisation of the graphic formulae might have done the trick instead, because as Ritter (2001: 44) notes:

> [G]raphical formulas so well accommodated the physical significance which, in time, accrued to them, that we automatically read physical relations into what were originally intended as (only) chemical relations.

In any event, this story illustrates two key claims. First, observable properties of models may be, but need not be, 'taken literally'. Whether they should be depends on the context. Second, reasoning in science often occurs with direct recourse to *observable* things and properties (whether by analogy or not)—even by manipulation of concrete models—and these provide a foundation for fostering understanding.

3. Electrons and Spin: Entities With Mixed Properties

The question now arises of how we should take talk of unobservable entities that are partially described in terms of (or assigned) observable properties, and partially described in terms of (or assigned) unobservable properties (and without recourse to effective analogies with observable systems). In order to answer this, I might have continued to consider chemical models of molecules and bonds in further depth. After all, Pauling (1970: 999) suggests that: 'bonds are theoretical constructs, idealizations, which have aided chemists during the past one hundred years in developing the convenient and extremely valuable classical structure theory of organic chemistry.'[29] However, bonds are a complex case in so far as they putatively involve links between other things, namely atoms, allegedly composed of further things, such as electrons, that might themselves be partially described in terms of unobservable properties. Thus I will instead discuss electrons, which are putatively fundamental.

Accept for present purposes that mass and charge are observable (e.g. when understood as dispositional properties to attract and repel).[30] What of spin? Is it an observable property? Here's a passage from a contemporary textbook:

> *The Electron Is Not Spinning*
>
> Although the concept of a spinning electron is conceptually useful, it should not be taken literally. The spin of the Earth is a mechanical rotation. On the other hand, electron spin is a purely quantum effect that gives the electron an angular momentum as if it were physically spinning.
>
> (Serway and Jewett 2013: 1315)

The claim here is that the electron isn't really spinning, although it *does* have an intrinsic angular momentum. But since angular momentum is classically defined in terms of movement about a point, one is left to wonder what is being discussed, or what *ontological* claim, if any, is being made here. The point is evident from considering the special case of a body in uniform circular motion. Its angular momentum is its linear momentum multiplied by the radius of the circle in which it moves: $L = mvr$.

Hence angular momentum is to momentum as torque is to force: both are moments, in the language of physics, in so far as they involve distances. But what is the *distance* we're supposed to be using in the case of the electron's 'angular momentum'?

Perhaps this is why looking a little further back, to the classic Landau and Lifshitz (1977: 198), one finds a subtly different claim:

> This property of elementary particles is peculiar to quantum theory . . . and therefore has in principle no classical interpretation. . . . In particular, it would be wholly meaningless to imagine the 'intrinsic' angular momentum of an elementary particle as being the result of its rotation 'about its own axis.'

Does 'intrinsic' appear in inverted commas for a reason? A full answer requires a historical treatment of how and why spin was introduced, which I'll postpone until Chapter 4. For the present, I will focus on the theoretical angle to answering this question.

Mathematically speaking, quantum mechanics involves wavefunctions and operators that may be applied to those wavefunctions. How to physically interpret this mathematical machinery is the controversial part. Indeed, most textbook presentations of quantum mechanics begin with contentious statements concerning 'observables', 'measurements', and so forth. To avoid falling into this trap, let's begin by noting merely that (1) a specific wavefunction is associated with any given entity (e.g. free electron) or system (e.g. free atom) treated by quantum mechanics, and that (2) the operators reveal information about the entity or system associated with the wavefunction when they are applied to the wavefunction. All standard interpretations of quantum mechanics agree on this.

The following may be added without taking a strong interpretative stance. First, each operator is *nominally* associated with a classical concept. There is a position operator, a momentum operator, and indeed a spin operator. Second, wavefunctions represent probability amplitudes. A probability amplitude is a complex number, which may be converted into a probability density function by multiplying it with its complex conjugate.[31] So probabilities and related averages that *nominally* involve classical concepts, but concern quantum entities or systems, may be calculated by employing operators, wavefunctions, and other simple mathematical techniques.

Here's a brief illustration of the kinds of calculation available. Let $\Psi(x)$ be the wavefunction associated with an entity, and $\hat{x}$ represent the 'position' operator. The average value we'd expect for 'position', $\langle x \rangle$, over an ensemble of entities associated with said wavefunction, is:

$$\langle x \rangle = \int_{-\infty}^{+\infty} \Psi(x)^{*} \, \hat{x} \Psi(x) \, \mathrm{d}x$$

This is a special case, which is easy to describe, because applying the 'position' operator to a function merely involves multiplying that function by 'position'. That's to say, $\hat{x}\Psi(x) = x\Psi(x)$, and it's easy to see that the equation involves integrating over a probability density function for x multiplied by x. To derive the momentum average, one needs to use the momentum operator instead, and so forth.

This brings us to interpretation proper. What do the aforementioned probabilities (and derivative averages) represent? Let's continue to consider only position, and assume, for simplicity's sake, that the unobservable things governed by quantum mechanics *sometimes* have positions (in a classical sense). Let's also assume that all position measurements are faithful, such that when an unobservable thing is measured as being in position r, it is in position r at that instant. We are left with the following question. Do these unobservable things *always* have positions?

To answer in the affirmative is to deny that the wavefunction associated with a thing (or a system) provides a *complete* description thereof. It is to deny, in Bohm's (1952: 169) words, that: 'The wave function with its probability interpretation determines the most complete possible specification of the state of an individual system.' But this is compatible with thinking that all the aspects not described by the wavefunction are *hidden*, in the sense that we are unable to determine them:

> other equally consistent interpretations . . . involve additional elements or parameters permitting a detailed causal and continuous description of all processes . . . 'hidden' variables. As a matter of fact, whenever we have previously had recourse to statistical theories, we have always ultimately found that the laws governing the individual members of a statistical ensemble could be expressed in terms of just such hidden variables. . . . Perhaps then, our present quantum-mechanical averages are similarly a manifestation of hidden variables.
>
> (Bohm 1952: 168)

Consider statistical mechanics as an illustration.[32] The temperature of a gas is a function of the (average) velocities of its atomic or molecular constituents. These velocities are 'hidden' in Bohm's sense. Probabilities are used because they're hidden, in a way that will become clearer in short order.

Let's consider a more concrete measurement situation, namely the famous two-slits experiment, for illustrative purposes. As Young showed in 1803, light from a single source generates an interference pattern when it passes through two 'slits' (which were holes in the original experiment). 'The most beautiful [thought] experiment of all time' (Crease 2002: 19) shows that a similar pattern can be formed on a screen by firing one electron at a time at two slits, and that this effect ceases when one of the slits is closed. Thus, no interference *between* electrons is responsible for the effect. It is often claimed that Feynman (1965: §1.1) originally proposed this thought experiment; Bach et al. (2013), for example, describe it as 'Feynman's thought experiment'. However, as we'll see, Bohm (1952: 173–175) discussed it at considerable length. It has now arguably been performed, moreover, despite Feynman's misgivings about this possibility.[33]

So we have a scenario where there's one electron, and at most we can predict where it will *probably* end up. But we have two key interpretative options. The first is to say that the probabilities are present in the world *in the scenario in isolation*—are 'single case propensities'—and reflect indeterminism in the fundamental laws of nature.[34] The idea, in short, is that identical initial conditions may lead to different outcomes. The second is to say that the probabilities are not present in the world in the scenario in isolation, but are (1) employed due to our ignorance of the position of the electron in individual experiments, and (2) represent how the ensemble of *empirically (or observably) indistinguishable* experiments end. The compatibility of (1) with (2) is sometimes missed. This is partly because two different interpretations of probability are potentially involved. Begin by considering a single experiment (*qua* firing of one electron). We use the probabilities because we can't *distinguish* the initial conditions in that experiment token from those in other tokens of the same type. So we might, for instance, use the principle of indifference to assign equal probabilities to each possible initial position of the electron (and derive probabilities for the outcome on that basis). We'd be using an *information-based*—a subjective, logical, or objective Bayesian—interpretation of probability. However, now consider the entire class of (empirically indistinguishable) experiments of the same type. Note that this contains one experiment for each possible initial position value. (The only thing that varies, between experiments of this type, are position values.) Then imagine randomly selecting and performing an experiment from this class. This procedure—associated with the experiment type—has a *long-run* propensity to produce results with specific frequencies. In this precise sense, we may say that the experiment type involves a world-based probability. Bohm (1952: 171) puts it so:

> In the usual interpretation . . . the need for a probability description is regarded as inherent in the very structure of matter . . . whereas in our interpretation, it arises . . . because from one measurement to the next, we cannot in practice predict or control the precise location of a particle, as a result of corresponding unpredictable and uncontrollable disturbances introduced by the measuring apparatus. Thus, in our interpretation, the use of a statistical ensemble is (as in

the case of classical statistical mechanics) only a practical necessity, and not a reflection of an inherent limitation on the precision with which it is correct for us to conceive of the variables defining the state of the system.

The moral of this story is that the choice between these key alternatives is underdetermined on the basis of empirical factors alone. As Cushing (1994: 214–215) puts it:

> In the end, we may be left with an essential underdetermination in our most fundamental physical theory and this issues in an observational equivalence between indeterminism and determinism in the basic structure of our world. . . . One possible conclusion is that, as a pragmatic matter, we can simply choose, from among the consistent, empirically adequate theories on offer at any time, that one which allows us best to 'understand' the phenomena of nature, while not confusing this practical virtue with any argument for the 'truth' or faithfulness of the representation of the story thus chosen. Successful theories can prove to be poor guides in providing deep ontological lessons about the nature of physical reality.

I agree with Cushing that underdetermination is present in this case.[35] And I view his 'possible conclusion' as roughly correct.[36] However, I did not cover this ground in order to point this out (although it foreshadows the discussion in the next chapter). After all, the presence of more than one possible interpretation doesn't entail that any such interpretation should be construed non-literally (as opposed to false). And as I already noted, in Section 2, Bohr's interpretation of quantum mechanics was explicitly non-literal.[37] On his view, we might *think* of the electron as a wave at some points and as a particle at others, in considering the two slits experiment, but shouldn't impute wave-particle character to the electron. That's because being a wave-particle is a simple contradiction in terms (and hence not an observable property). In short, conflicting views of the same entity or arrangement *for different purposes* are allowable, for Bohr, because:

> in our description of nature the purpose is not to disclose the real essence of phenomena but only to track down as far as possible relations between the multifold aspects of our experience.
>
> (Bohr 1934: 18)

So why have I discussed the introduction of hidden variables? To illustrate that when one starts down the path of interpreting the discourse in quantum mechanics in a realist fashion—when one starts to interpret the theory as telling a story that goes beyond the measurements, to impute classical properties to some of the unobservable things governed by it, perhaps to treat the wavefunction as real (in some sense), and so forth— one eventually comes to a point where one must interpret some of the discourse non-literally. And spin is a case in point, when it comes to Bohmian mechanics. I quote from Goldstein (2013):

> [A]fter the discovery of quantum mechanics it quickly became almost universal to speak of an experiment associated with an operator *A* . . . as a *measurement of the observable A*—as if the operator somehow corresponded to a property of the system that the experiment in some sense measures. . . . The case of spin illustrates nicely . . . some of the difficulties that the naïve realism about operators mentioned above causes. . . . Bohmian mechanics makes sense for particles with spin. . . . When such particles are suitably directed toward Stern-Gerlach magnets, they emerge moving in more or less a discrete set of directions. . . . This occurs because the Stern-Gerlach magnets are so designed and oriented that a wave packet (a localized wave function with reasonably well defined velocity) directed towards the magnet will, by virtue of the Schrödinger evolution, separate into distinct packets—corresponding to the spin components of the wave function and moving in the discrete set of directions. *The particle itself, depending upon its initial position, ends up in one of the packets moving in one of the directions.*
>
> The probability distribution for the result of such a Stern-Gerlach experiment can be conveniently expressed in terms of the quantum mechanical spin operators. . . . From a Bohmian perspective there is no hint of paradox in any of this—unless we assume that the spin operators correspond to genuine properties of the particles. [emphasis added]
>
> (Goldstein 2013)[38]

Hence one shouldn't take the talk of spin (even *qua* intrinsic property of the electron) literally, while taking the other elements of the Bohmian picture (such as electron position and momentum) literally. To do so would be inconsistent.

I haven't considered—and will not, for obvious reasons, consider— all possible interpretations of quantum mechanics. But I have given an indication of the difficulties in pursuing a fully literal interpretation of quantum mechanical discourse, and that there's a tension between taking talk of position literally while also taking talk of spin literally. And one can hardly take talk of spin literally—in any classically respectable sense (e.g. as an intrinsic angular momentum)—while *not* taking talk of position literally. There's no sense in talking about a spinning object with no position. There's also no sense in talking about a moment like angular momentum, as mentioned earlier, without a distance.

We now come to the question that inspired this section. Should we take talk of electrons literally? The property instrumentalist's answer lies in the negative. But talk of *some tiny thing possessing discrete mass and charge* may be taken seriously; and a property instrumentalist might accept that we have isolated such an entity, while believing that the talk of 'spin' (or 'intrinsic angular momentum') is just a way to account for its behaviour in some circumstances. It is crucial to note, however, that spin need not be understood to be a fiction merely for explaining how *observables* interrelate. Rather, it may be taken to be a fiction for predicting—and furnishing us with an understanding of—how those unobservable things 'similar to electrons' behave.

For clarity, I should emphasise that talk of things 'having an intrinsic unobservable property of a kind with which we're not acquainted' may be construed literally in principle, according to property instrumentalism. However, ascribing such a property is quite different from ascribing spin *qua* self-rotation or spin *qua* intrinsic angular momentum.

4. On the Contemporary Applicability of Property Instrumentalism

Thus far, I have outlined a form of semantic instrumentalism that is more moderate and defensible than its predecessors. I have already shown that it is reasonable in principle, with reference to the concept of spin. But the question remains as to whether it is reasonable in practice. Next, I will use a brief example from contemporary physics in order to argue that it is.

This concerns the virtual particles of quantum field theory, which *ex hypothesi* violate the relativistic equation for the total energy of an entity, $E^2 = p^2c^2 + m_0^2c^4$ (where p is momentum and m_0 is rest mass), due to considerations relating to uncertainty, and may even possess negative energies and impart negative momentum. A property instrumentalist, however, will likely construe talk of negative energy (and therefore virtual particles) as non-literal in this context. The argument would run that the concept of energy is derived from, and explicable in terms of, observables such as velocity and mass. And in classical contexts, when employing the equation $E_k = pv/2$, negative kinetic energy is precluded (although it is possible to speak of negative potential energies since the zero-point may be arbitrarily defined). What matters is the magnitude of the velocity and the mass of the object under consideration (which can be related to how much of it there is on some conceptions, as Jammer [1961] discusses). The property instrumentalist would not object to understanding the notion of energy literally when this is extended into the unobservable realm in a way that is compatible with how it is applied in the observable realm. But this is inconsistent with positing negative energies.[39]

The difference between the semantic realist and the property instrumentalist is therefore brought into perspective by the Casimir effect—i.e., that two parallel uncharged metallic plates brought very close together in a vacuum experience an attractive force—which has recently been demonstrated in a number of experiments.[40] See Figure 2.4. A full-blooded semantic realist will take this as evidence for the existence of virtual particles. She will accept the explanation that only virtual particles with wavelengths that fit a whole number of times into the gap between two plates will appear there, whereas virtual particles with a broader range of wavelengths will appear outside. A property instrumentalist will not accept this explanation (as true or approximately true as opposed to convenient), provided she does not take talk of virtual photons literally.

Moreover, some physicists agree with the stance on virtual particles suggested by property instrumentalism. For example, the website of the SLAC National Accelerator Laboratory, which is operated by Stanford University, states:

> Virtual particles are a language invented by physicists in order to talk about processes in terms of the Feynman diagrams. These diagrams are a shorthand for a calculation that gives the probability of the process. . . . Particle physicists talk about these processes as if the particles exchanged in the intermediate stages of a diagram are actually there, but they are really only part of a quantum probability calculation.[41]

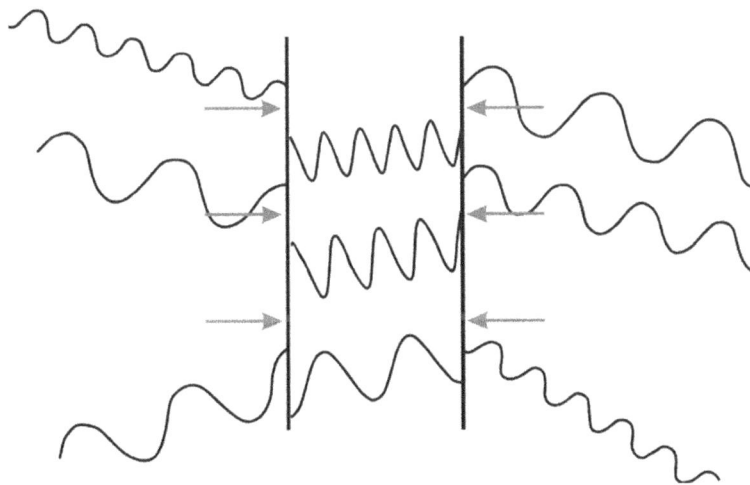

Figure 2.4 Depiction of the Casimir Effect in terms of Virtual Particles[42]

Virtual particles are an interesting example because almost anyone adopting property instrumentalism would agree that talk of these is not to be construed literally. And other examples from physics—such as tachyons (which possess imaginary mass)—are possible. It should not be forgotten, however, that property instrumentalists may legitimately disagree with one another about what is an observable property (or allowable extrapolation thereof) or an observable object.

5. Conclusion

In this chapter, I have explained and advanced property instrumentalism, which is the second component of cognitive instrumentalism. I have argued that talk of unobservable objects should be taken literally, at most, *only* to the extent that those objects are assigned properties, or described in terms of analogies involving other things, with which we are experientially (or otherwise) acquainted. I paid special attention to the role of analogies and models involving familiar observable things in science. I presented a case study, on chemical notation, to illustrate how such models can be construed non-literally but come to be considered literally, and can serve as useful vehicles for reasoning about their target systems.

I closed by considering microscopic physics. First, with a focus on quantum mechanics and the notion of spin, I discussed how talk of entities with mixed properties, observable and unobservable, should be understood according to property instrumentalism. Second, with a focus on quantum field theory and virtual photons, I showed how property instrumentalism bears on more recent physics.

Notes

1 This chapter contains material from: 'The Instrumentalist's New Clothes', *Philosophy of Science* 78(5), 1200–1211 (2011); and 'Models in Biology and Physics: What's the Difference?', *Foundations of Science* 14(4), 281–294 (2009).

2 So construed, traditional semantic instrumentalism is compatible with the view that sentences ostensibly referring to unobservable entities are truth evaluable, when they are properly interpreted; hence, it does not entail Bergmann's (1943: 339) view that: 'Theoretical or existential constructs . . . are fictitious or, as I should prefer to say, [purely] calculational things'. The 'standard empiricist construal of scientific theories' discussed by Hempel (2001: 209), which was developed by Carnap (1956), proceeds along these lines:

> According to the standard conception, a theory can be analytically divided into two constituent classes of sentences. . . . The sentences of the first class are viewed simply as a set of sentential formulas containing certain uninterpreted extralogical constants, namely, the theoretical terms. Let us assume these formulas to be axiomatized; the result will be an uninterpreted axiomatized formal system. The standard construal conceives the first set as such as system; it is sometimes called the theoretical calculus,

C. The sentences of the second class are viewed as affording interpretations of theoretical expressions in a vocabulary whose terms have fully determinate and clearly understood empirical meanings; they are said to constitute the set R of rules of correspondence, or of interpretative sentences, for the calculus C.

This position is instrumentalist in character because theories can be understood to lack non-empirical content. Nonetheless, Carnap advocated a neutral line on the realism debate, because he took this to rest on a pseudo-question about 'reality' (understood in an inappropriate metaphysical sense). Shortly before his death, he wrote: 'I believe that the question should not be discussed in the form: "Are theoretical entities real?" but rather in the form: "Shall we prefer a language of physics (and of science in general) that contains theoretical terms, or a language without such terms?" From this point of view the question becomes one of preference and practical decision' (Carnap 1974: 256). For more detail on Carnap's views, see Psillos (1999: ch. 3) and Friedman (2011). I take it to be uncontroversial that science is better off with a language containing such terms, even if they are eliminable in principle (as illustrated by Craig's [1953] theorem). The reason is adduced by Hempel (1958: 87): 'as far as suitability for inductive systematization, along with economy and heuristic fertility, are considered essential characteristics of a scientific theory, theoretical terms cannot be replaced without serious loss by formulations in terms of observables only'.

3 Sankey (2008) is a case in point. Even Psillos (1999) devotes only a little over twenty per cent of his monograph to tackling it (and what he calls 'reductive empiricism', which is a form of semantic instrumentalism in my terminology), precisely because his 'main focus' is 'the current debates over realism'.

4 I discuss the paper in which Maxwell makes this erroneous claim when I cover the observable-unobservable distinction in Chapter 6. For present purposes, note that Maxwell (1962: 13) admits: 'it would . . . be fatuous to minimize the importance of the observation base, for it is absolutely necessary as a confirmation base for statements which do refer to entities which are unobservable at a given time'. And consider how this fits with the suggestion that 'we "directly" observe many kinds of so-called theoretical things. . . . [I]f we were carrying a heavy suitcase in a changing gravitational field, we could observe the changes of the $G\mu\nu$ of the metric tensor' (Maxwell 1962: 14). If we allow such 'observations' in to the base—I maintain that we cannot observe such changes, although we can *interpret* our observations in such a way—then it is difficult to see how radical theory change can occur in a principled way via conflict with experience. Arguably what would allow us to make such observation statements is our sensory experience of the force acting on the suitcase (which would result in *more basic* observation statements considering how heavy it felt at various points). Those *more basic* observation statements are compatible with the falsity of our theories concerning gravity, and therefore with the fact that there are no changes of the metric tensor to observe. That is, even if they are *also* theory laden. And almost any new theory proposed would be dismissed if it did not account for statements about those sensations, although this would be methodologically absurd if they were no more basic, pragmatically if not epistemically, than statements about metric tensors.

5 The position advocated here is a rather like a *partial* reductive empiricism, in the terminology of Psillos (1999).

6 In the words of Faye (2008): 'Bohr denied that classical concepts could be used to attribute properties to a physical world in-itself behind the phenomena, i.e. properties different from those being observed.' See also Folse

(1994), who suggests, by contrast, that Bohr is better understood as leaning towards realism rather than anti-realism. My treatment of Bohr's work in atomic physics, in Chapter 4, supports Faye's view.

7 See also McMullin (1984: 14): 'The realist claim is that the scientist is discovering the structures of the world; it is not required in addition that these structures be imaginable in the categories of the macroworld.'

8 This is not to concede that there is not a solid ontological line, on the epistemic account of vagueness championed by Williamson (1994).

9 The property instrumentalist accepts that we 'are able to state in the observational vocabulary . . . that there are unobservable entities, and to some extent, what they are like' (van Fraassen 1980: 54) and thereby avoids van Fraassen's line of criticism against some instrumentalist, or reductive empiricist, projects.

10 Similarly, on the question of the existence of unobservables, McMichael (1985: 267) notes: 'we observe that all bits of matter, save perhaps for those at the limits of vision, can be divided into smaller parts. . . . It is true that in the case of objects at the limit of vision, division does not yield *perceptible* parts, but that is *not* to observe that the objects have not been divided into *imperceptible* parts.'

11 See also Mach (1893: 587): 'Science acts and acts only in the domain of *uncompleted* experience. Exemplars of such branches of science are the theories of elasticity and of the conduction of heat, both of which ascribe to the smallest particles of matter only such properties as observation supplies in the study of the larger portions.'

12 Note also that the gravitational constant, G, can be defined as $1/4\pi\Gamma$, where Γ is an analogue to the permittivity of free space. The factor of 4π doesn't appear in the Newton's law of gravitation, although it does appear in Coulomb's law of electrostatics, because of different conventions in defining gravitational and electric flux.

13 For objections, see Smith (1977), Prior, Pargetter and Jackson (1982), Johnston (1992), Martin (1994), Armstrong (1997), Lewis (1997), Bird (1998), and Choi (2003).

14 I don't presume that all metaphysical theories are empirically equivalent. But many are. A case in point concerns whether properties are universals or particulars. (Or indeed whether they're multi-grade universals. See MacBride [2005].)

15 The popular names for various atomic models almost always involve analogies too. More examples are these: billiard ball (Dalton), plum pudding (Thomson), and planetary (Rutherford). However, such names don't always correspond to analogies explicitly invoked by the scientists. For instance, 'plum pudding' is not Thomson's phrase. As we will see in Chapter 4, it is also misleading. Electrons are not static in Thomson's model. Plums in pudding are.

16 As van Fraassen (2008: 31) emphasises: 'what determines the representation relationship . . . can at best be a relation of *what is in it* to factors *neither in the artifact itself nor in what is being represented* . . . [such as] *use, practice, and context*'.

17 For more on the Phillips model, see Phillips (1950), Morgan and Boumans (2004), and Moghadam and Carter (2006).

18 Thanks to John Christie for drawing this passage to my attention. Even earlier still, Kepler wrote: 'I cherish more than anything else the Analogies, my most trustworthy masters. They know all the secrets of Nature' (Polya 1973). As Gentner et al. (1997: 439) show, Kepler employed many analogies in his work: 'In pursuit of a causal model of the planetary system, Kepler analogized sun and planet to sailors in a current, magnets, a balance beam, and light, to

name only some of the more prominent analogies.' Gentner et al. also relate their findings to those of Dunbar (1995). They write: '[Dunbar finds that] the process of working through an analogy contributes to scientists' on-line creative thinking, and [this] . . . lends force to Kepler's introspection that analogy furthered—and perhaps even engendered—his theories. Another possible parallel stems from Dunbar's observation that the heterogeneity of the laboratory group contributes to creativity. Dunbar speculates that this is true in part because group heterogeneity increases the range of different analogues that can be brought to bear. The idea that a stock of different analogues is conducive to creative thought accords with our conclusions concerning Kepler' (Gentner et al. 1997: 446).

19 This quotation is from a note that was published only after Fitzgerald's death, which explains the grammatical errors.

20 The importance of 'superficial' analogies is also supported by Dunbar's (2002: 158–159) study on how scientists use analogies in formulating hypotheses, in which he concludes that:

> [W]hile scientists use many analogies that share superficial features between the source and target, they can and do produce analogies that are based on deeper structural features. Thus, analogy-use appears to be flexible, and to change with the goals of the analogizer.

On a similar note, from a philosophical perspective, Bailer-Jones (2009: 192) writes:

> Keeping in mind the function of a model will help to decide in which respects the model must be truthful in order to meet its function. What are the possible functions of models? . . . to predict something, to explain, or to instruct . . . for being creative, for exploring a new path of thinking.

21 It doesn't involve merely the set of properties that the modeller does not expect to do any such work. Hence, 'not intending to suggest' is a further slip on Hesse's part. Hesse's 'not' is in the wrong place.

22 Harriot's thoughts about sphere stacking also encouraged him to think of atoms as spheres. See Shirley (1983: 242) and Hales (2000).

23 Crum Brown began to devise the system a few years earlier, in his M.D. thesis. It was based on some earlier notational ideas of Couper and Dalton. See Ritter (2001) for more.

24 Crum Brown (1864: 708) states that his notion was devised not to 'indicate the physical, but merely the chemical position of the atoms'. As Meinel (2004: 249) explains, 'chemical' in this passage means something akin to 'functional'.

25 See, however, the fascinating diagram in the thesis—reproduced by Ritter (2001: 37)—in which ethanol is represented entirely in terms of H symbols (in circles). That's to say, two proximal H symbols replace the O symbol, and four proximal H symbols replace the C symbols. Rocke (1983) understands this to represent a hypothesis about the subatomic cause of valence.

26 The origin of the set depicted isn't clear. But the label inside the lid reads 'GLYPTIC FORMULAE'. Meinel (2004: 253) writes:

> In May 1867 the journal *The Laboratory* inserted a brief editorial note advertising a set of Hofmannian "glyptic formulae". . . . Made by a certain Mr. Blakeman of Gray's Inn Road, London . . . the set was praised for the striking constructions that could be assembled from it. . . . It is difficult to tell how such model kits were used. The scarcity of examples in museum collections argues for their being used up in classroom teaching

and thrown away afterwards: transient objects that could be made out of cheap materials by even the least skilled laboratory assistant. Accordingly, most surviving molecular model sets have no manufacturer's name, although size, design, and colour codes are very similar.

27 This had been Hofmann's view for some time, as he had previously devised another concrete model, where wire frames were used to hold cubes representing different atoms or atomic groups. See Hofmann (1862). Meinel (2004: 246–247) writes of this model:

> Convinced as he was that symbolic notations in chemistry were purely formal tools that did not immediately correspond to reality, this approach explicitly avoided the question of truth. Consequently, Hofmann's type moulds and atomic cubes were not meant to represent the physical arrangement of the atoms. They rather supplied a pattern according to which the chemical operations of elimination and substitution could be classified and analogies found.

28 In his M.D. thesis, but not in his paper of 1864, Crum Brown referred to the lines connecting atomic symbols—which were then dotted—as 'lines of force'. So there is perhaps a case to be made that he, too, took these to represent real connections between atoms (at least at that stage in his career). The 'force' language marks the model out as mechanical, even if static, at the bare minimum.

29 The debate on bonds continues, with competing 'structural' and 'energetic' conceptions being offered by those with realist inclinations. See, for example, Hendry (2008) and Weisberg (2008).

30 Mass (as opposed to weight) is an interesting case. It is plausibly observable on several of the understandings covered by Jammer (1961).

31 Wavefunctions may be spinors, say when spin is involved, but I leave this to one side in the interests of brevity. Nothing depends on this in the present context.

32 Bohm's view is similar to the earlier 'wave-pilot' view of De Broglie, and 'the De Broglie-Bohm view' is often written of. However, Bohm's interpretation is different from De Broglie's in significant respects. For example, the wavefunction is said to be in configuration space (in $3n$ dimensions, where n is the number of objects it is associated with) in the former, but to be a wave in 3D space in the latter. See Cushing (1994: 148–149 & 251–252, f. 33). Whether the wavefunction should be understood to be real, on a Bohmian view, is a matter of dispute; see Dürr et al. (1997).

33 See Tonomura et al. (1989), which reports results of an experiment using an electron biprism in place of slits, and Bach et al. (2013), which reports an alleged realisation of the thought experiment.

34 For more on single case propensities, and a comparison with long run propensities, see Gillies (2000: ch. 6, ch. 7) and Rowbottom (2015: ch. 8). Here, I will put aside the concern that single case propensities might not be identifiable with probabilities as standardly defined (e.g. in the Kolmogorov and Popper axioms). See Lyon (2014).

35 Cushing (1994: 213) also points out how liable we are to impute properties of *mathematical* laws or equations to the world:

> It is in . . . laws or equations, which represent our world, that we discover and explore the property of determinism. Once we have satisfied ourselves of the accuracy and reliability of these mathematical laws, we then are willing to transfer (perhaps unconsciously) the generic features of these

laws . . . to the physical world itself. We go beyond a mere instrumentalist view of the laws of science . . . and take them realistically as literally true representations of the world. What is our warrant for this transfer?

Cushing doubts we have ever had much warrant, due to the existence of systems whose behaviour we could never predict (such as the double pendulum); he describes such transfers, aptly, as 'faith' based.

36 It's only roughly correct because competing theories need not be 'empirically adequate' to be empirically as useful, as such, and because there are trade-offs to be struck between theoretical virtues (sometimes on a context-by-context basis). Moreover, we should remain mindful of the distinct role of models.

37 As noted earlier, there is considerable controversy about Bohr's views. Camilleri and Schlosshauer (2015: 76), for example, have recently argued that:

> The real question for Bohr was not what kind of reality is described by quantum mechanics, but rather "what kind of knowledge can be obtained concerning objects" by means of experiment. Thus, for Bohr, the epistemological problem of quantum mechanics was not how we should interpret the formalism, but rather how *experimental knowledge* of quantum objects is possible.

There is, nevertheless, reasonable agreement that Bohr's thoughts about the importance of classical concepts were epistemological in character. And I would maintain—in line with the account of his views defended by Faye (1991)—that we would do well not to ignore passages in Bohr's writing such as:

> [T]he appropriate physical interpretation of the symbolic quantum-mechanical formalism amounts only to predictions, of determinate or statistical character, pertaining to individual phenomena appearing under conditions defined by classical physical concepts.
>
> (Bohr 1958: 64)

> Strictly speaking, the mathematical formalism of quantum mechanics and electrodynamics merely offers rules of calculation for the deduction of expectations about observation obtained under well-defined experimental conditions specified by classical physical concepts.
>
> (Bohr 1963: 60)

What's ultimately important, however, is merely that some physicists *have* defended positions in line with the view that I attribute to Bohr, often under the (highly misleading) name of 'the Copenhagen interpretation'.

38 For more on naïve realism about operators, see Daumer et al. (1997).

39 If desired, the issue of negative energy need not be raised. Consider virtual photons, for example; it is possible for these to 'carry momentum but no energy' (Williams 1991: 15). Classically, this makes no sense in so far as the two physical concepts are connected. If momentum is positive then mass and speed must be positive. And if mass and speed are positive, then kinetic energy must be positive.

40 See Lamoreaux (1997), Mohideen and Anushree (1998), and Bressi et al. (2002).

41 URL (accessed on 15/01/10): http://www2.slac.stanford.edu/vvc/theory/virtual.html

42 Thanks to Pete Edwards for permission to use this diagram.

3   Unconceived Alternatives and the Contingency of Science's Content[1]

In the previous two chapters, I discussed scientific progress and scientific discourse. In the former, I argued that science progresses primarily by increasing the extent to which it saves the phenomena and furnishes us with understanding of how said phenomena interrelate. In the latter, I argued that scientific discourse concerning unobservable entities should only be taken literally in so far as it involves observable properties or analogies with observable things. The presence of non-literal discourse—concerning, for instance, spin and virtual photons—is to be explained by the role it plays in promoting and achieving progress of the aforementioned kind. The first two elements of cognitive instrumentalism cohere.

In the present chapter, I further advance the view that that science *is* a cognitive instrument—for saving the phenomena and providing understanding thereof—by criticising the dominant alternative realist view, that science is instead a reliable means by which to discover what the unobservable world is like. More precisely, I argue that it is reasonable to harbour strong doubt that theories in contemporary science concerning the unobservable are approximately true, or are likely to be replaced by theories with greater truthlikeness. The argument here is independent of those that have gone before.[2] So it is effective, for instance, even if it is granted that all discourse concerning unobservable things should be treated literally. The core premise is that our conceptual space is significantly limited at any point in time.

Before I begin, however, an important remark is in order. Other realist views that are less strong in character—such as entity realism, structural realism, and semi-realism—may also be viable alternatives to cognitive instrumentalism. However, I postpone discussion of these other positions, which are less popular, until Chapter 6. This said, the argument developed in this chapter potentially bears on these weaker forms of realism. For example, if it's granted that unconceived (seriously competing) theories often say different things about the structure of the unobservable world, then the argument from unconceived alternative theories bears on (most variants of) structural realism.

1. The Significance of Unconceived Theories

I will begin by presenting what I take to be the strongest version of the argument from unconceived alternative theories, the original version of which was presented by Stanford (2006). I will then present some new arguments with similar targets; several concern how values (or estimated values) of prior probabilities may change unexpectedly and unpredictably.

Scientific realists typically think there is a correlation between confirmation and truthlikeness. For instance, Maxwell (1962: 18) writes that: 'well-confirmed theories are conjunctions of well-confirmed, genuine statements and . . . the entities to which they refer in all probability exist'; Boyd (1980: 613) claims that: 'scientific theories embody the sorts of propositions whose (approximate) truth can be confirmed by the ordinary experimental methods which scientists employ'; and Psillos (1999: 70) states that: 'insofar as scientific theories are well confirmed, it is rational to believe in the existence of the entities they posit.'[3] The significance of unconceived alternative theories may therefore be illustrated by appeal to confirmation theory. Assume, for the sake of illustration, that the confirmation of a hypothesis, h, is equal to its conditional probability given some evidence, e, in the presence of some background information (or 'knowledge'), b.[4] (This assumption is not necessary.)[5] Then the confirmation value can be calculated by Bayes's theorem (in the form used by Salmon 1990a):

$$P(h,eb) = \frac{P(h,b)P(e,hb)}{P(e,b)}$$

$P(e, b)$ decomposes into $P(h, b)P(e, hb) + P(\sim h, b)P(e, \sim hb)$. And $P(\sim h, b)P(e, \sim hb)$ in turn decomposes into $P(h_1, b)P(e, h_1b) +, \ldots, + P(h_n, b)$ $P(e, h_nb)$, where the set of possible alternatives to h is $\{h_1, \ldots, h_n\}$. Note also that h could be replaced with 'approximately h', if desired, without affecting any of the subsequent arguments. One technical way to do this would be to let h represent a disjunction of appropriately similar theories. They would be 'appropriately similar' in the following sense: if one disjunct were true, then all the other disjuncts would be approximately true (and no further competing hypotheses, outside this set, would be approximately true).[6]

Let's now think in terms of the subjective interpretation of probability, which is the most popular among contemporary confirmation theorists (and Bayesians in particular), for illustrative purposes.[7] A theory h is only highly confirmed by e provided that $P(\sim h, b)P(e, \sim hb)$ is low relative to $P(h, b)P(e, hb)$. Moreover, if we want to allow that scientific experiments can highly confirm theories that are strongly doubted, initially, we should not stipulate that $P(h, b)$ must be high. It follows—since $P(\sim h, b) = 1 - P(h, b)$—that we should also not stipulate that $P(\sim h, b)$ must be low.

Here's a historical scenario that shows we should allow for such cases. In 1819, a committee of the *Académie des Sciences* met to judge to whom to award a *Grand Prix* for work on diffraction. One of the papers they considered, by Fresnel, championed a wave theory of light. However, many of the panel members were convinced that light is corpuscular. Poisson was one such panel member, and he sought to refute Fresnel's theory by deriving from it a consequence that he took to be obviously false. This consequence was that a bright spot appears in the middle of the shadow cast by an opaque disc (or sphere). However, another panel member (and future prime minister of France), Arago, performed the experiment and found the bright spot. As a result, Fresnel's paper won the prize.

So let h be Fresnel's wave theory, and e be (a report of) the bright spot appearing in Arago's experiment. This is a case where $P(e, b)$ and $P(h, b)$ were low for many of the panel members, and $P(h, eb)$ was judged to be very high.[8] So $P(e, \sim hb)$ is a key term which must have been low.[9]

Naturally, it is *possible* for one's subjective probability in $P(e, \sim hb)$ to be low, and for $P(\sim h, b)P(e, \sim hb)$ to be low as a result. However, said probability would become dramatically higher, if a new serious alternative predicting e (or predicting e to an appropriately similar extent to h) became apparent. (A 'serious' alternative in this subjective context means an alternative that the individual would take seriously, hence its prior probability would be reasonably high.) Then the confirmation value of h would become considerably lower. Thus, confirmation values may lower considerably, *as a result of a newly conceived alternative theory with the appropriate properties.*

The subjective interpretation was used only for illustrative purposes. Doubtless many scientific realists hold that confirmation rests on more than psychology (even of a mob variety); they prefer a non-subjective account of confirmation, based on a logical, objective Bayesian, or perhaps even a frequency or propensity view of probability. But even if one adopts such an account, on which *actual* confirmation values never fluctuate, one should nevertheless concede that our *estimates* of those values may fluctuate as a result of our changing information about the alternatives to h. As Salmon (1990b: 329) puts it:

> What is the likelihood of any given piece of evidence with respect to the catchall? This question strikes me as utterly intractable; to answer it we would have to predict the future course of the history of science.[10]

Salmon's solution to the problem is to consider only the (positively) *conceived* alternatives to h.[11] And on the basis of these, we can calculate the confirmation of h *relative* to the conceived alternatives. However, relative confirmation has no obvious connection to truthlikeness, even on the assumption that absolute confirmation (in some non-subjective sense) does indicate truthlikeness (or probable truthlikeness, or whatever surrogate one prefers). Hence, there are no confirmational grounds for thinking that h is truth-like unless there are grounds for thinking that there are no serious unconceived alternatives to h. And this grants the bold assumption that the possible is a subset of the conceivable.

Now grant that there have been many serious unconceived alternative theories in the past, as Stanford (2006) argues.[12] The significance of this, for the tenability of scientific realism, *does not depend on any inductive inference* from the past to the present (and future), although Stanford (2006) does make such an inference in his original version of the argument from unconceived alternatives.[13] Rather, it poses a challenge for the realist who claims that contemporary theories are typically approximately true, provided that they are well-confirmed.[14] This is as follows.

Why be confident that the confirmation value of any given theory (on a subjective view), or the estimate thereof (on an objective view), would not change drastically if all the unconceived alternatives were appreciated? What licences inferring absolute confirmation values from relative confirmation values? If the realist cannot answer satisfactorily, it is reasonable to deny that contemporary theories are approximately true, and that future theories will probably be closer to the truth than current theories.[15] As we will see in what follows, the force of this challenge may be strengthened by appeal to unconceived observations. We will also see how even our estimates of *relative* confirmation can be unstable and/ or incorrect, for independent reasons to do with unconceived models, unconceived experiments, and the like.

We will now begin to consider these different kinds of unconceived entities, many of which are connected in interesting and subtle ways. The findings in several of the different sections may also be connected; for example, prediction and explanation are two sides of the same coin, if Hempel's (1965) symmetry thesis holds.

2. Unconceived Observations

Put on hold the idea that some experiments—types or tokens—might not be conceived of, despite being conceivable, at any given stage in science. Why else might observations—and related observation statements—fail to be conceived of? One possible scenario is as follows. The observations in question are theory-laden, and the theory (or set of theories) necessary to conceive of them is unconceived.

Imagine the following hypothetical scenario. It's 1850, and archaeologists are exploring the remains of an ancient civilisation, which spanned the Iberian Peninsula. A striking feature of the civilisation is the art, which involves many depictions of ducks, drawn in the same style. Murals of ducks are found in ruins of (buildings thought for independent reasons to be) temples, and pictures of ducks are found buried with the dead. The archaeologists take this to be evidence that ducks had some kind of special religious or spiritual significance in the civilisation. They are somewhat surprised not to have found many remains of ducks in their archaeological work. But they suspect that the animals were treated as sacred, and allowed to roam free.

A few years later, however, there is a remarkable new find. Elsewhere in Iberia, the preserved remains of a previously unknown animal—a lagomorph—are discovered. Scientists decide to call it 'the rabbit'. (In this scenario, no-one has before encountered rabbits because they were wiped out by a remarkably infectious virus—somewhat similar to our very own myxoma—before they spread beyond Iberia.)[16] And it is not long before a young archaeologist hypothesises that the aforementioned art depicts rabbits. He publishes his magnum opus on the ancient civilisation, and the centrality of this noble beast therein. He goes on to have a glittering career, as one of the leading lights of archaeology.

The moral of the story is as follows. Singular observation statements concerning ducks may now be replaced with singular observation statements concerning rabbits.[17] So from one point of view, the nature of the evidence is unstable. For one theory does not explain the presence of duck art, whereas the other does not explain the presence of rabbit art. From another point of view, the evidence remains the same—the pictures on the murals, and so on, are unaltered—but a different interpretation thereof is available. We do not need to decide which view is better, for present purposes.[18] Either way—whether e changes to e^* on some of the serious alternatives to h, or there are some serious alternatives to h that predict e because they involve unexpected interpretations thereof—there are highly unpredictable routes by which confirmation values can change.[19]

Here's a brief illustration. It would be odd to insist that the prior probability attached to 'Ducks are depicted' *should* be higher than that attached to 'Rabbits are depicted' (relative to background information that both kinds of animal were around at the time). Moreover, both theories save the (relevant) phenomena. Hence, if we let the duck theory be represented by h and the rabbit theory be represented by h^*, we may say—as a matter of fact on a non-subjective interpretation of probability, and for some reasonable people on *subjective* view of probability—either that: (1) $P(h, b) \approx P(h^*, b)$ and $P(e, hb) = P(e, h^*b)$; or (2) $P(h, b) \approx P(h^*, b)$ and $P(e, hb) = P(e^*, h^*b)$, where any suitable theory should account for *either e or e* (and e and e^* are mutually exclusive).

I used a hypothetical example to keep matters simple. Most relevant cases in the history of science are rather more complicated. But here's one brief example. In medieval Europe, phenomena such as comets were typically viewed, like meteors, as being sublunary. And this was largely due to Aristotle's influence; generation and corruption are distinctly terrestrial, as opposed to celestial, on his view. However, Tycho Brahe saw (or interpreted) the 1572 supernova as a superlunary event, and took this to provide evidence against the aforementioned Aristotelian hypothesis (among other things). But Brahe only saw (or interpreted) the supernova as superlunary on the basis of other observations, and other theories that he held (which many previous astronomers had not). He measured the parallax of the supernova, and judged this to be too low for it to be superlunary; in short, it appeared to move like any other star. (Analogously, in the previous story, a new type of animal was discovered. It was only *counted* as a new type of animal on the basis of scientific theories not mentioned in the story.) So Brahe's observations did not rule out the (rather *ad hoc*) hypothesis that the supernova was sublunary, but moved in such a way as to be consistent with his parallax measurements.

Admittedly, this example might be critiqued as follows. The alternative—that the supernova was superlunary—had been conceived of, but not treated seriously. First, however, it may be retorted that scientists don't have a perfect memory concerning past alternatives. Hence, a community may not have conceived of an alternative even when that alternative has, historically, been conceived of. Second, even were it to be accepted, this criticism doesn't do much work. For it remains apparent that assessments of the relative merits of theories are highly dependent on what's accepted as background information—whether or not other background information sets have been conceived of—and that this is liable to change in ways that aren't predictable.

To summarise, the possibility of unconceived observations of the kind discussed so far is significant in *raising* the plausibility of the claim that there may be *serious* unconceived alternative theories. We will discuss unconceived observations due to unconceived experiments, rather than unconceived theories, later.

In closing this section, I should mention unconceived observations of a final kind. These involve new and unanticipated phenomena which may be seen *without* theoretical changes. The bright spot discussed in the first section is an interesting case in point, because it was noted by Deslisle (1715) and Maraldi (1723) around a century before Fresnel's paper, although the panel judging the paper were unaware of this. Sometimes, moreover, such new and unanticipated phenomena are encountered in non-experimental contexts. The appearances of newly discovered plants and animals are cases in point, and the discovery of chemosynthetic ecosystems is especially striking.[20] Similarly, no-one conceived of the appearance of the brontosaurus until bones from the beast were discovered. If we were to encounter a well-preserved brontosaurus, moreover, we might still be surprised by its appearance. The order of such appearances may be contingent, and affect the direction of science.

3. Unconceived Models and Unconceived Predictions

Models are necessary in science because theories alone often lack appropriate predictive force. Recall the discussion of pendulum motion in classical mechanics in Chapter 1. An early model was the simple pendulum; the mass of the rod bearing the bob is ignored, as is friction, and a small angle of swing is assumed (such that the sine of the angle is approximately equal to the angle). The movement is taken to occur only in two dimensions. And so on. But the adequacy of classical mechanics to deal with real pendulum motion was unclear initially, in so far as more sophisticated models were yet to be conceived of. It is also easy to create only slightly more complicated systems, in terms of component parts, which are much harder to deal with. Consider the double pendulum, which comprises one pendulum connected to the bottom of another. The motion of such an arrangement is chaotic; a small change in initial conditions can have a profound effect on the future development of the system.

Tractable models with fewer idealisations were developed only slowly, over a period of time. And much of the predictive power of classical mechanics was unclear for over a hundred years after Newton. Lagrangian and Hamiltonian mechanics were vital for some applications. (Lagrangian mechanics, for example, employs scalars rather than vectors; among other things, this makes it easier to perform co-ordinate transformations.) Such *reformulations* of classical mechanics—and hence models employing them—were not readily apparent. In part, this supports Butterfield's (2004) view that such reformulations fall between the levels of 'laws of nature' and 'models'.

Why does this matter for confirmation? In essence, unconceived models may be responsible for unconceived *predictions*, and the resources of a theory may fail to be apparent—and be underestimated (or even overestimated)[21]—as a result. A semi-formal illustration follows. (Think now in non-subjective terms, for simplicity's sake.) Let e represent the total body of available evidence that a theory in mechanics is expected to account for. And let h and h^* represent the two available theories in mechanics (i.e., the only two conceived theories that have not been shown to predict $\sim e$ in conjunction with b). $C(h^*, e, b)$ may be much higher than $C(h, e, b)$ because the estimate of $P(e, hb)$ is much lower than it should be. And it may be much lower than it should be due purely to unconceived models based on h (or unconceived reformulations of h).[22] For example, h and b might entail e, whereas h^* and b might not. However, only the following might have been shown: h^* and b entail e^*, and h and b entail $e^\dagger$, where e^* and $e^\dagger$ are each proper subsets of e, and $e^\dagger$ is a proper subset of e^*.[23]

In summary, even judgements of *comparative* confirmation depend on judgements about the predictive power of theories, and such judgements are contingent on the available, and hence *conceived*, models. So why be confident that there are no unconceived models that would affect (esti-mated) confirmation values? This is a further challenge to realism.

Consider also models of a non-theoretical variety. Some of the most striking and potentially distinctive are model organisms, which Keller (2002: 51–52) describes as follows:

> [U]nlike mechanical and mathematical models (and this may be the crucial point), model organisms are exemplars or natural models—not artifactually constructed but selected from nature's very own workshop. . . .
>
> Model organisms represent in an entirely different sense of the word than do models in the physical sciences: they stand not for a class of phenomena but for a class of organisms. As such, they are more closely akin to political representatives, and, in fact, are employed in a similar fashion—as a way of inferring the properties (or behaviour) of other organisms. It is just for this reason that bio-logical modelling has sometimes been described as proceeding "by homology" rather than "by analogy". . . .
>
> The primary criterion for the selection of a model organism is only rarely its simplicity—the principal criterion for a model in the physical sciences. Far more important is the experimental accessibil-ity endowed by particular features (such as size, visibility, reproduc-tive rate).

However, the differences between models and modelling in the physical sciences and biology are not as dramatic as Keller suggests. First, not all biological models are natural; the fruit flies (*drosophila melanogaster*) used in genetics, for example, have been engineered.[24] Second, concrete models feature in physics. In the words of Galison (1997: 693): 'Such models have long histories; one thinks for example of nineteenth century British electricians, as they put together pulleys, springs, and rotors to recreate the relations embodied in electromagnetism.' Think also of the antikythera mechanism, which modelled the motion of celestial bodies—see De Solla Price (1974)—in the ancient world. This analogue com-puter animated theories of how such bodies move relative to one another (and of how their relative positions affect other phenomena, such as the phases of the moon). In one part of the mechanism, as Freeth (2009: 82) explains:

> the two gears turned on slightly different axes. . . . Thus, if one gear turned at a constant rate, the other gear's rate kept varying between slightly faster and slightly slower. . . . [T]he varying rotation rate is precisely what is needed to calculate the moon's motion accord-ing to the most advanced astronomical theory of the second cen-tury B.C., the one often attributed to Hipparchos of Rhodes. . . . In

Hipparchos's model, the moon moved at a constant rate around a circle whose center itself moved around a circle at a constant rate—a fairly good approximation of the moon's apparent motion.

We have also seen in Chapter 2 that concrete models appear in chemistry, and even economics; think back to the discussion of Hofmann's glyptic formulae, which 'were realised by concrete three-dimensional structures of croquet balls and connecting arms' (Hendry 2012: 294–295), and to the discussion of Phillips's hydraulic model of the economy.[25] Gelfert (2016: 115–127) also discusses several other interesting examples (of what he instead calls *material* models), such as the Watson-Crick model of DNA and the wax models of Wilhelm His.

Third, modelling by homology has been important outside biology. Consider, for example, the '*homology transformations* of stellar models' employed by Gamow (1939). To put matters simply, two stars are genuinely homologous (in a classical sense) if their relative mass distributions are equal; and, as a result, their pressure and density distributions are related in a simple way. Think of a star as a sphere. Now consider that sphere as a series of concentric shells. The radius of each is a particular fraction of the radius of the whole star ('its fractional radius'), and the mass inside it is a particular fraction of the mass of the whole star ('its fractional mass'). For two homologous stars, fractional radii and fractional masses map onto each other. Unsurprisingly, few star pairs are genuinely homologous. However, some are approximately so, and hence the transformations have some useful applications. As Kippenhahn and Weigert (1990: 191) put it:

> the conditions . . . are so severe that real stars will scarcely match them. There are a few cases, however, for which homology relations offer a rough, but helpful, indication for interpreting or predicting the numerical solutions.

All this shows that concrete models are important in science as a whole, and function as means for reasoning in a variety of ways. And there may be unconceived concrete models. For example, no *undiscovered* organism has been conceived of as a model. Moreover, it is possible not to conceive of using a known organism as a model, through ignorance about some of its properties. It might be thought to be less 'experimentally accessible' than it actually is. Or, to foreshadow some of the later discussion, it might be 'experimentally inaccessible' only in virtue of a lack of appropriate instruments. Moreover, as we've already seen, model organisms need not be wild in origin. Rather, they may be synthetic like laboratory *drosophila melanogaster*. Hence there are many unconceived forms of synthesised organism, in so far as constructing such organisms is a complex process, and the results are often tricky to predict.[26]

Model organisms share other things in common with the (abstract and concrete) models discussed in the previous two chapters. In the words of Ankeny (2009: 202), they involve or embody 'certain simplifying assumptions and idealizations', and these are necessary because 'If research were always done on the actual, naturally occurring biological entities, it would be difficult to make testable predictions or to link results obtained from systems at different levels (e.g., the genetic and the phenotypic).' Ankeny (2009: 202) also echoes some of my earlier thoughts about understanding, which are illustrated with historical examples in the next chapter, when she writes:

> Putting . . . points across in the form of a story allows a better understanding than would simple provision of a complete description of the actual situation.

This concludes the discussion of unconceived models (of both concrete and abstract varieties) and unconceived predictions. In summary, models connect theories to the world, in so far as they are (almost always) necessary to derive predictions from theories. So conceiving of a new model might result in unexpected, and even unconceived, predictions. Those predictions might, in turn, affect the confirmation of the theory. (And the scope of the theory might also be reassessed as a result.)

4. Unconceived Explanations

Put aside the previous worries about models, and imagine, for the time being, that observation statements can typically be derived directly from theories (without even the need for reformulations).[27] Assume also a syntactic view of theories; assume that theories may be understood as collections of propositions or sentences. Now, for the sake of exposition, we can use Hempel's (1965) deductive-nomological account of explanation. On this view, an explanans must be true, entail the explanandum, contain a general law statement ('theory'), and have empirical content. (This is the basic picture, although some small refinements may be added. For example, it should be stipulated that no *proper* subset of the propositions in the explanans should entail the explanandum.) Hence, the explanans for 'The pen hit the ground one second after it was dropped' might once have been thought to involve Newton's law of gravitation and Newton's second law of motion, the mass of the Earth, the mass of the pen, the distance between the centre of mass of the Earth and the centre of mass of the pen, and the distance between the pen and the ground.

Because the explanans should be true, however, we should take into account the rotation of Earth, use general relativity instead of Newtonian mechanics (assuming the former is true), and so forth.[28] Thus, it becomes extremely difficult, at any point in time, to distinguish between an *actual* and a *potential* explanation. So let's just discuss *potential* explanatory power in what follows. The *potential* explanatory power of a theory (or bundle of theories) depends only on which *known* observation statements it entails (when conjoined with true statements of initial conditions). Think of it as what the theory would explain *if it were true*. The truth status of said theory (or bundle of theories) is irrelevant.

Now think about how we measure potential explanatory power more carefully. As noted previously, we require *true statements of initial conditions*. But even granting that we can determine whether any statement of initial conditions is true, when it's considered, a problem remains. For in some cases, *we may simply fail to conceive of the initial conditions*.

Consider, for example, the history of the study of the tides. One threshold moment was Newton's treatment in the *Principia*. But this only showed that some aspects of the tides could (potentially) be explained. As David Cartwright (1999: 2) notes:

> From time to time a new idea has arisen to cast fresh light on the subject. While such events have spurred some to follow up the new ideas and their implications, they have also had a negative effect by appearing superficially to solve all the outstanding problems. Newton's gravitational theory of tides. . . [potentially] explained so many previously misunderstood phenomena that British scientists in the 18th century saw little point in pursuing the subject further.

The *superficial* explanatory power noticed by Cartwright arises, to some extent, because of the unconceived initial conditions in (and concerning) our seas and oceans, which are highly complex. So in effect, beliefs that the periods of the tides in any specific area could be (potentially) explained by Newtonian mechanics, in the eighteenth century, were based on a dubious extrapolation from successes in some contexts to future successes in others. It was *not* just a matter of thinking that the values of variables of known types, such as ocean floor topography and coastal geography, were relevant to saving the phenomena. It was, moreover, a matter of thinking that all the *relevant* types of variables had been conceived of. But they hadn't. The discovery of Kelvin waves, for instance, came considerably later. And this sort of pattern has been repeated throughout the history of research into the tides, according to Cartwright (1999: 1):

> [E]very improvement in accuracy of measurement and prediction has led to further fundamental research into previously hidden details.[29]

In essence, the point here is that *judgements of explanatory power are liable to change considerably, just as judgements of predictive power are*, as the limits of the conceived expand. And judgements of the relative merit of theories, on the basis of estimated explanatory power, are liable to change as a result.

Incidentally, in using the D-N account of explanation I have also advanced a further argument that judgements of predictive power may be highly error prone, provided that there are relevant cases where explanation is symmetrical with prediction (i.e., where the explanations are potential predictions). This does not require that explanation is symmetrical with prediction in general.

5. Unconceived Experiments, Methods, and Instruments

This brings us to experiments. Even given a theory and models that render it predictive in a domain of interest, all the possible experiments involving it are not manifest. Partly, this is due to instruments and methods that have not been conceived of. Think of the role played by the torsion balance in Cavendish's (1798) ingenious measurement of the mean density of the Earth. Newton could not have conceived of such an experiment without having first, at the bare minimum, conceived of such a balance. But the balance was invented long after his death, in the late eighteenth century. (Michell, a close friend of Cavendish's, devised the balance and conceived of 'the Cavendish experiment' long before it was performed. However, knowledge of the balance only became common after it was independently invented and used by Coulomb in 1777.)[30] New ways to measure the gravitational constant (which may be easily calculated from the aforementioned density), reliant on more sophisticated instruments, continue to be devised; the most recent experiment, performed by Rosi et al. (2014), achieves an astounding reduction in experimental error (if one accepts the premises). This fancy work, involving laser-cooled atoms and quantum interferometry, is a far cry from anything dreamt of by Newton or Cavendish.

This is far from the whole story, in so far as instruments need not feature at all. To design an effective experiment, or an experiment that is possible to perform given funding constraints, may require a great deal of ingenuity. Consider blind and double blind experiments, which were possible—and possible to *positively* conceive of—for centuries. Nonetheless, the first recorded example occurred in the late eighteenth century, when King Louis XVI appointed commissioners to investigate animal magnetism.[31]

Why does this matter? Scientists' assessments of their theories depend on the evidence at their disposal. (Such evidence also affects their assessments of the attractiveness of the research programmes involving said theories.) And the available experiments delimit the available evidence. Hence, which theories are more confirmed/corroborated, and therefore whether progress towards truth occurs, is (sometimes) contingent on which experiments are conceived of.

The significance of unconceived experiments is greater still if novel predictions have more power to confirm than accommodations, as argued by philosophers such as Maher (1988) and Douglas and Magnus (2013).[32] For the extent to which we can make novel predictions is contingent upon the new experiment types—and not merely new experiment tokens—that we can conceive of. Indeed, some theories have plausibly suffered, in comparison to their counterparts, precisely because they made no new predictions. Consider again Bohm's 'interpretation' of quantum mechanics, which is a different quantum mechanical theory from those seriously considered beforehand, from a realist perspective, due to its distinctive claims about the unobservable, most notably that particles have definite positions at all points in time and that their states evolve deterministically. The mere fact that Bohm's 'interpretation' appeared after the Copenhagen 'interpretation' would make it less confirmed by the evidence, given that the two theories are (apparently) empirically equivalent and the latter was used to predict some of the evidence that the former was not (and the converse does not hold).[33] Thus its *contingent* fate as a marginal (or 'sidelined') theory—as illustrated by Cushing (1994)—was appropriate, provided that its prior probability was (and remained, as background information changed) no higher than that of its rival. The relevance of such contingency for realism is summed up nicely by Cushing (1990: 192–193), with reference to the way that methods may change:

> [E]ven in situations in which there are empirically equivalent theories . . . who gets to the top of the hill first holds the high ground and must be dislodged (if required, not otherwise). There is often nothing compelling (or unique) about the direction or particular *course* of evolution of (the history of) scientific theories. This tells against scientific realism, there being nothing necessary about the particular theory or picture of the world we finally accept. Once a successful, stable theory has gained sway, there is a temptation to imbue it with a (scientific) realist interpretation, both to increase our understanding (picturability) of its laws and to strengthen its claim to necessity or uniqueness. We have a natural inclination to seek a simple, necessary and compelling picture of the way the world must be. An invariant set of methodological rules would increase our epistemic security in these final products of science. We may *want* these, but a close attention to actual scientific practice does not allow (at least some of) us to harbor these fictions.

And maybe there is an experiment, as yet unconceived, that would discriminate between the Copenhagen and Bohmian views? It would be the height of arrogance to be certain that there is not, in so far as the predictions we can make from the theories depend, as is evident from some of the formal representations we considered above, on *background information* (including *auxiliary hypotheses*). And why should we even think that there is *probably* not any such experiment? What does the realist know about how background information will probably change, in the future, which licences that inference? Again, this is a challenge. It is not a rhetorical question.

Consider also one final sense in which unconceived experiments can result in alterations of confirmation/corroboration values, on views which link such values closely to hypothesis testing. There is an intuitive sense—which might be made more precise in a variety of formal fashions—in which some tests are more severe than others. And for some philosophers of science, how strongly a theory is to be preferred is a function of how well it has been tested. But then, of course, the fates of theories depend on the experimental tests conceived of. For example, Popper (1959: 418) writes: '$C(h, e)$ can be interpreted as degree of corroboration only if e is *a report on the severest tests we have been able to design*.' So in one reading, which is explored in detail in Rowbottom (2008a), merely designing (*qua* conceiving of) a new experiment—which can be performed in practice, and not merely in principle, perhaps—is sufficient to render current corroboration values irrelevant. That's because one can't have a report on the severest tests one has designed unless one has also performed said tests. Consider, in this regard, the remarkable experiment performed on Gravity Probe B, involving the motion of a gyroscope orbiting the Earth; see Everitt et al. (2011). This probe was launched over 40 years after Schiff (1960) proposed such a test for the geodetic effect, noting that 'experimental difficulties . . . are greatly reduced if the gyroscope does not have to be supported against gravity . . . experiments of this type might be more easily performed in a satellite'. Similarly, in the modern day, tests for gravity waves in orbit have been proposed, and may soon be funded after the remarkable claim of Abbott et al. (2016: 061102–1) to have detected a gravity wave with 'a false alarm rate . . . less than 1 event per 203,000 years'.[34]

6. Unconceived Values (or Theoretical Virtues)

A final item is values *qua* theoretical virtues. Consider, for example, Kuhn's (1977: 321) list thereof: 'accuracy, consistency, scope, simplicity and fruitfulness'. The importance of some of these has been assumed in the previous discussion. For example, I discussed how the limits of what we've conceived might adversely affect our estimates of accuracy and scope, and touched on how unconceived theories may be simpler than, despite being otherwise as virtuous as, their conceived counterparts. Indeed, one rough way to present the standard argument from unconceived alternatives is this: 'Unconceived theories may be—or are, or often are—more virtuous than those we've conceived of.'

How we *rank or weigh* the virtues, even assuming that we agree on what they are, will affect the values assigned to priors, such as $P(h,b)$.[35] For example, you and I might prefer different theories simply because I think that simplicity is more valuable than scope, whereas you think that scope is more valuable than simplicity. (This is irrespective of our individual stances on the realism debate. We may agree on what the theoretical virtues are, but disagree on whether they are pragmatic or epistemic in character.) Here, however, I'm concerned with whether there are virtues that we've not conceived of, and in whether conceptions of virtues change in interesting ways over time. From a realist perspective, for example, are there indicators of truthlikeness that we have not yet conceived of (and therefore failed to recognise)? Is there any principled way to show that the probability of such unconceived theoretical virtues is low?

Kuhn (1977: 335) concludes, on the basis of his limited sample from the history of science, that: 'If the list of relevant values is kept short . . . and if their specification is left vague, then such values as accuracy, scope and fruitfulness are permanent attributes of science'. In order to stack the argumentative deck (concerning unconceived values) in the realist's favour, let's grant this. Let's grant also that Kuhn's list of values is exhaustive, so that we do not have to invoke mysterious undiscovered values. A question still remains as to whether understanding of those values can change over time. For example, might one woman's simplicity be another woman's complexity?

The point is not merely that simplicity may be sub-divided into syntactic ('elegance') and ontological ('parsimony') varieties (among others, perhaps). Rather, the notion is that what counts as simple, even within such sub-divisions, may nevertheless be a matter of legitimate dispute. Consider elegance in the case of astronomical models of the solar system. The findings of Kuhn (1957) support the conclusion that reasonable disputes may occur. For example, should Tusi couples be used in place of Ptolemaic equants? The equant was introduced in Ptolemy's *Almagest*. It is a point from which the motion of the centre of an epicycle, along its deferent, appears uniform; see Neugebauer (1975: 155) and Evans (1984). The Tusi couple is named after Nasir al-Din al-Tusi, who devised it in his 1247 commentary on the *Almagest*. It is a mathematical device whereby a circle rotates inside, while remaining in contact with the circumference of, another circle of twice the diameter.[36]

Let h be a theory (or model) involving the former, and h^* be a theory (or model) involving the latter. It may be the case that $P(h, b)$ is higher than $P(h^*, b)$ and $P(h, b^*)$ is lower than $P(h^*, b^*)$, where b and b^* represent different background assumptions.

Similar concerns arise concerning consistency (in so far as h may be consistent with other scientific theories relative to b, but not b^*) and fruitfulness (which is notoriously difficult to measure, in any event), but I will not press the point here. The nature of this kind of challenge to realism is already evident.

7. Conclusion

Grant the (implausible) thesis that the possible is a subset of the conceivable in practice. What's conceived is nonetheless limited, for a variety of reasons; limitations on time and material resources, contingencies about where attention is directed, and so forth. The tenability of scientific realism of a convergent variety depends on those limits being less significant, over time. And the tenability of the view that contemporary ('well-confirmed') theories are approximately true, even in what they say about the unobservable, relies on those limits being insignificant in a *remarkable number* of (rather diverse) respects.

Given the absence of effective arguments that those limits are insignificant in these respects, *agnosticism* about the truthlikeness of contemporary theories (and the future direction of science with regard to truth) is prudent. Even if you think that the probability that each kind of unconceived alternative is significant is low, moreover, the probability that they are significant in combination may be high; the situation is *additive*.[37] And if the burden of proof lies on those who advance the theses under attack here for independent reasons—as Belot (In Press) argues[38]—then it is reasonable to conclude that they are false.

Now consider the findings in the previous chapters in combination with those here. If there's a significant chance that there are several aspects of the world we can't grasp, as suggested in Chapter 2, then this adds force to the view that we shouldn't go beyond agnosticism—or non-realism—in the epistemic dimension. But this is not a troubling conclusion if the core value of science is instrumental, as argued in Chapter 1, or a puzzling conclusion if empirical success does not require truth-like theories, as illustrated at several points in the book. So the elements of cognitive instrumentalism combine into a cohesive image of science.

Notes

1 This chapter contains content from 'Extending the Argument from Unconceived Alternatives: Observations, Models, Predictions, Explanations, Methods, Instruments, and Values', *Synthese* (2016) (https://doi.org/10.1007/s11229-016-1132-y).

2 Dependent arguments are also possible, but I take these to have been touched upon sufficiently already. For example, if unobservable entities possess many unobervable properties, then our grasp of their natures will be limited significantly (even if we can successfully refer to them).

3 On the next page, Psillos (1999: 71) also denies, by implication, that it's rational to fail to believe in those theories (when one's aware of them, and their high confirmation values, presumably), when he writes: 'successful scientific theories *should* be accepted as true (or, better, near true) descriptions

of the world in both its observable and its unobservable aspects' [emphasis added].

4 I prefer 'background information' to 'background knowledge' for reasons explained in Rowbottom (2014b). See also Williamson (2015) for an alternative view of *b* (and *e*).

5 The important requirement, which will become evident in the discussion that follows, is that the confirmation (or corroboration) value depends on $P(e, \sim hb)$. This holds for all standard confirmation (or corroboration) functions, such as those championed by Popper (1983), Milne (1996), and Huber (2008).

6 Considering the situation in this more sophisticated way doesn't help, but instead appears to hinder, the realist case. For example, there's no reason to suppose we'll typically be aware of all the members of the set, even if we have identified some such members. And different members might have subtly different consequences. Thus values for $P(e, h)$ could be sensitive to conceiving of new members of the set.

7 I prefer a group level interpretation, which can cover research groups or scientists in specific fields, in this context. Such interpretations are articulated and defended in Gillies (2000) and Rowbottom (2013a, 2015: ch. 6).

8 For more analysis of this episode, see Worrall (1989a) and Rowbottom (2011d).

9 In case this is unclear: note that $P(e, hb)$ was unity, such that $P(h, eb)$ was $P(h, b)/P(e, b)$; the posterior probability could have been high only if the denominator and numerator had similar values. Moreover, for $P(e, b)$ to be small, $P(h, b)P(e, hb) + P(\sim h, b)P(e, \sim hb)$, into which it decomposes, must be small. And we know $P(h, b)$ was small, and therefore that $P(\sim h, b)$ was large. Thus, $P(e, \sim hb)$ was small.

10 Salmon adopts a frequency-based view, but might have done better, given the problems identified by Hájek (1996, 2009), to adopt a long-run propensity view. See Gillies (2000: ch. 5–7) and Rowbottom (2015: ch. 7–8).

11 The form of conceivability under discussion here is 'positive' in the sense discussed by Chalmers (2002: 153), although he primarily discusses situations: 'to positively conceive of a situation is to imagine (in some sense) a specific configuration of objects and properties'.

12 I will not recapitulate his historical case. Bohm's interpretation of quantum mechanics—which was a genuine alternative to any forerunner from a realist perspective, in so far as it says different things about the unobservable realm—is an example that was raised in the last chapter.

13 Stanford (2001: S9) writes: 'the history of scientific inquiry offers a straightforward inductive rationale for thinking that there typically are alternatives to our best theories equally well-confirmed by the evidence, even when we are unable to conceive of them at the time'.

14 The historical cases studied by Laudan (1981) may also be understood simply to cast doubt on the putative (probabilistic) connection between empirical success and successful reference of central theoretical terms (and/or approximate truth). No appeal to induction is then needed.

15 So construed, the problem of unconceived alternatives also presents a challenge to *some* possible forms of anti-realism. For example, it casts doubt on the view that contemporary theories save the phenomena, or some proper subset thereof, in the most elegant (or more generally, virtuous) way possible. However, no theses such as these are involved in cognitive instrumentalism.

16 Pedantic readers might think that hares are sufficiently similar for the pictures to be seen as hare-ducks. But imagine, if you will, that the whole *leporidae* family was wiped out by the virus, which badly affected hares as well as rabbits (unlike myxomatosis).

17 They may also be replaced with observation statements concerning duck-rabbits, and this is potentially important from the point of view of scientific method. I will avoid discussing this possibility, however, in order to streamline the discussion.

18 Perhaps there are two different senses of 'evidence'—one subjective/intersubjective, and the other objective—employed here. That is, unless the subjective/intersubjective evidence is taken to be non-propositional. My view is that there are *some* situations where the evidence itself changes, although this hypothetical scenario may not be one of them.

19 This may be conceded without adopting any form of extreme relativism, or collapsing the distinction between fact and theory. One need not go as far as Feyerabend (1958). The point can hold even if one simply agrees with Harré (1959: 43): 'that only some descriptive statements involve terms whose meanings depend partly on theory.' In any event, realists have used theory-ladenness as an argument for the view that the line between the observable and the unobservable can shift; see, for instance, Maxwell (1962) and the discussion in Chapter 6.

20 Thanks to Ian Kidd for raising this example. As Van Dover (2000: xvii) explains:

> Deep-sea hydrothermal vents and their attendant faunas were discovered in 1977. While the hot-water springs were predicted to occur at seafloor spreading centers, no one expected to find them colonized by exotic invertebrate faunas. Accustomed to a view of the deep sea as a food-limited environment, the puzzle of how lush communities could be maintained provoked biologists into a flurry of research activity. . . . Based on collections from the early expeditions to hydrothermal vents in 1979 and 1982, investigators identified the significance of chemoautotrophic primary production in these systems.

See also www.divediscover.whoi.edu/ventcd/vent_discovery/—a Woods Hole Oceanographic Institute webpage that contains interviews with many of those involved in the discovery and early investigations.

21 See the discussion of expectations concerning Newtonian mechanics and the tides in the next section.

22 Tractability is an important issue, which is bound up with the talk of models and reformulations, as we saw in Chapter 1. Here's a further example from David Cartwright (1999: 2):

> Solution of Laplace's tidal equations, even in seas of idealized shape, taxed mathematicians for well over a century until the advent of modern computers. Even then, some decades were to elapse before computers were large [sic] enough to represent the global ocean in sufficient detail, and techniques had improved sufficiently to give realistic results.

23 The presentation here is simplified, because the hypotheses in conjunction with the models will entail the evidence. Thus if the models are sunk into background information, for instance, then b will *change* as the new models are developed. This doesn't affect the thrust of the argument.

24 As Kohler (1994: 53) puts it:

> [T]he little fly was . . . redesigned and reconstructed into a new kind of laboratory instrument, a living analogue of microscopes, galvanometers, or analytical reagents. . . . "Standard" drosophilas were constructed from stocks that produced recombination data conforming most closely to Mendelian theory. . . . Their chromosomes were a bricolage. . . . The purposes

and key concepts of mapping were thus built physically into domesticated drosophilas: what better indication of their artifactual nature?

More generally, as Ankeny (2009: 200–201) explains:

> [M]odel organisms, particularly successful ones, are often extremely distant from those that could be easily found outside the laboratory and in nature, which might be considered by some to be the 'real' organisms. However, this means that there are often clear limits that are built into the experimental systems . . . that often are noneliminable, hence restricting researchers from making direct inferences back to the naturally occurring organisms.

25 For more on Phillips's model, see Phillips (1950), Morgan and Boumans (2004), and Moghadam and Carter (2006).
26 See, for instance, Ankeny (2009: 196–197) on the parasite *mycoplasma genitalium*, which is relatively simple (in so far as it has a small genome) and lives exactly where you'd expect from the name.
27 As Frigg (2009) notes:

> In some cases the equations that form part of the model can be obtained from a general theory simply by specifying the relevant determinables in the general equations of the theory. . . . Hence the relation between theory and model is that between the general and the particular.

In the Newtonian model of the Sun-Earth system, for example:

> [T]he model-equation . . . is obtained from a general theory—Newtonian mechanics—by specifying the number of particles and their interaction. This equation specifies a model-structure, which is instantiated in the model-system.

(Frigg 2010: 133)

28 I here treat explanation as factive, in line with what many realists believe, in order to streamline the discussion.
29 Surprisingly, Cartwright (1999: 4) nevertheless endorses convergent realism at one level: 'the *global* aspects of tidal science . . . seem to have reached a state of near-culmination'.
30 See McCormmach (2012: §6.11.3) for more on the collaboration between Michell and Cavendish.
31 For a brief summary of the episode, and references to some of the relevant literature, see Kaptchuk (1998) and Best et al. (2003). See Kaptchuk (1998, n. 9) for a mention of some precursors.
32 The opposing view is defended by Harker (2008). For a nice summary of the historical views on this issue, see Musgrave (1974).
33 As noted by Faye (2008), 'Copenhagen interpretation' is really 'a label introduced . . . to identify . . . the common features behind the Bohr-Heisenberg interpretation'.
34 This false alarm rate is misleading, in so far as the calculation assumes that the data has been 'cleaned up' properly, and so forth. The paper is also remarkable in so far as it claims that the detected wave demonstrates 'the existence of binary stellar-mass black hole systems' (061102–1). However, any wave detected here could be the result of interference effects.
35 For further discussion of this phenomenon, with particular attention to the interpretation of probability, see Rowbottom (2011d: ch. 3).
36 Tusi originally described the couple differently, such that both circles move in opposite directions, but the resultant effect would be the same; see Kennedy (1966) for more.

37 Analogously, the probability that one of five horses wins a race may be greater than a half although the probability that each wins the race may be only slightly higher than a tenth.

38 Belot's case, made on the basis of geophysical (and astrophysical) examples, rests on considering claims concerning *states of affairs*:

> [I]n the case of gravimetry: corresponding to any given configuration of the gravitational field external to the Earth is a vast (infinite-dimensional) family of possible internal structures (some of which will differ very dramatically from one another). Travel-time tomography constitutes an interesting intermediate case: for some internal structures the Earth might have, underdetermination would evaporate in the limit of complete data; for others, even complete data would be consistent with a vast and various family of possibilities. . . .

> Very often, one is interested in systems with infinitely many degrees of freedom. In that case, a finite number of observations will never suffice to determine the state of the system. If one considers an infinite number of possible observations, the underdetermination may or may not disappear—this depends on the details of the theory determining the connection between states and data and on the space of states in question.

4 Historical Illuminations
1885–1930

A Briton wants emotion in his science, something to raise enthusiasm,
something with human interest. He is not content with dry catalogues.
—Fitzgerald (1896: 441)

[A]tomic physics . . . recommended itself only to the hardiest even among
the British.
—Heilbron (1977: 45)

In the previous three chapters, I presented and defended the core of cognitive instrumentalism; in doing so, I employed several examples from the
history of science. I will now discuss a key period of historical science—
some *historic* science—in greater depth. My main aims in doing so are: to
illustrate how cognitive instrumentalism fits actual scientific practice; to
adduce further historical evidence in favour of claims made in previous
chapters; and to use the historical examples to elaborate on some other
pertinent topics.

I will focus on atomic theory, and related developments, from 1885
to 1930. I take this period to be of special significance, in so far as the
realism debate is concerned, because it involves numerous developments
that realists are inclined to construe as great discoveries concerning the
unobservable world. These include the discovery of atoms and several of
their component parts.

I will begin by covering an interesting finding in spectroscopy, and
then say something about other significant developments in spectroscopy
beforehand. Subsequently, I will discuss several important experimental
findings and theoretical developments that followed. This will set the
scene for a discussion of the eventual positing of spin.

The treatment of the positing of spin ties up a loose end from the
second chapter; it provides further evidence that talk of 'spin' should
not be taken literally (because 'spin' is not an observable property). The
prior history of atomic theory provides an opportune illustration of how
empirical findings can outstrip theoretical findings—of how empirical

laws can be discovered in the absence of any theoretical counterparts—and of how models explicitly designed to be understood semi-literally may serve to unify and to provide an understanding of such empirical findings. As you will see, it is apt that I am British, *qua* cognitive instrumentalist, in so far as:

> Atomic models . . . were very much a British speciality. According to the Victorian tradition, models served heuristic purposes rather than representing some reality of nature. They were first of all mental illustrations, formulated mathematically and based on the established laws of mechanics and electrodynamics, if often supplemented with hypothetical forces. . . . [They were] not to be taken literally, but were seen as a method or picture that offered some insight into the inner workings of nature.
>
> (Kragh 2012: 11)

This fits with the way that molecular models were treated by chemists of the same period, as we saw in Chapter 2. Recall, for example, the billiard ball arrangement—the concrete model—used by Hofmann (1865). Meinel (2004: 245) explains the motivation for using this as follows:

> *Anschaulichkeit*, the ability to appeal to the mind's eye by transforming abstract notions into vivid mental images, was Hofmann's chief pedagogical method. Impressive demonstration rather than abstract reasoning was supposed to transmit scientific knowledge. In this way theoretical notions turned into mental images could be read as a language.

The Victorians recognised that a model need not be concrete to achieve this demonstrative feat. They saw that an abstract model could do the trick. What's more, they took such models to be indispensable for fostering understanding—building 'insight'—and hence conducting good science in many contexts.

1. Spectroscopic Beginnings: Zeeman, Balmer, and Rydberg

When some substances are exposed to magnetic fields, their spectra change. And when the magnetic fields are sufficiently strong, the spectral lines 'split'. The title of Zeeman (1897b)—'Doublets and Triplets in the Spectrum Produced by External Magnetic Forces'—describes the splitting effect nicely, which is typically referred to as the Zeeman effect.

When Zeeman found this effect, however, there was still no satisfactory theoretical explanation of the spectra of even simple atoms, such as hydrogen. The electron was only in the process of being discovered:

for instance, Zeeman (1897a) writes of: 'Lorentz's theory . . . that in
all bodies small electrically charged particles with a definite mass are
present, that all electric phenomena are dependent upon the configuration and motion of these "ions," and that light-vibrations are vibrations
of these ions.' (I shall adopt a realist mode of discourse, at most points,
to make this section more eloquent. So read me charitably. 'Spectra of
even simple atoms' may be understood to mean 'spectra of (empirically
distinguishable) substances that were composed of simple atoms according to the dominant theories at that time', 'discovered' may be read as
'widely believed to have been discovered', and so forth. Such long and
awkward constructions make for tiresome prose.[1]) As suggested by this
brief quotation, Zeeman played an important part in the discovery of the
electron, although its discovery is often attributed solely to Thomson. As
Arabatzis (2001: 188) puts it:

> Several physicists, theoreticians and experimentalists provided evi-
> dence that supported the electron hypothesis. The most that can be
> said about one of those, say Zeeman, is that his contribution to the
> acceptance of the electron hypothesis was significant.

In short order, I intend to go forwards in time from Zeeman's discovery, to discuss early atomic theories. But before doing so, I want to
say some more about the intellectual context. So let's rewind a little.
Our story begins with a 60-year-old schoolteacher and occasional university lecturer in geometry, a mathematician with rather eccentric
numerological interests, who set his mind to answering the following
question.[2] What's the mathematical relationship between hydrogen's
spectral lines? It didn't take this man, Balmer, long to find an empirical
formula. And remarkably, he knew only of the first four visible lines in
the spectrum—namely Hα, Hβ, Hγ, and Hδ—when he spotted the relationship. That is, if we take his report at face value. In his own words
(Balmer 1885: 81):

> The wavelengths of the first four hydrogen lines are obtained by mul-
> tiplying the fundamental number h = 3645.6 in succession by the
> coefficients 9/5; 4/3; 25/21; and 9/8. At first it appears that these
> four coefficients do not form a regular series; but if we multiply the
> numerators in the second and the fourth terms by 4 a consistent regu-
> larity is evident and the coefficients have for numerators the numbers
> $3^2, 4^2, 5^2, 6^2$ and for denominators a number that is less by 4. For sev-
> eral reasons it seems to me probable that the four coefficients which
> have just been given belong to two series, so that the second series
> includes the terms of the first series; hence I have finally arrived at
> the present formula for the coefficients in the more general form: $m^2/$
> $(m^2\text{-}n^2)$ in which m and n are whole numbers.

As Balmer (1885) goes on to show, existing measurements of five further lines, in the ultraviolet region, were in close agreement with this formula.[3] (The fifth and sixth lines are visible, the sixth only barely to me, despite their official classification as 'UV'.) In concluding, Balmer (1885) writes 'we may perhaps assume that the formula that holds for hydrogen is a special case of a more general formula'.

Act one, scene three. Enter Rydberg. He had been working on understanding the relationship between the spectral lines of the alkali (or 'group one') metals for several years, and was assisted by Balmer's formula. Just three years later, he devised a more general formula, relating wavelength (in a vacuum), λ, to a constant, R, and two further positive integer variables, n and m:

$$\frac{1}{\lambda} = R\left(\frac{1}{n^2} - \frac{1}{m^2}\right) \text{ where } m > n \tag{0}$$

So Balmer's formula only covered one *series* in the spectrum of hydrogen, where n is equal to two. The existence of other series, in which none of the lines fall in the visible part of the spectrum, is predicted by Rydberg's equation in so far as n may be higher than two.[4] It wasn't until the early twentieth century, however, that lines outside the Balmer series were generated in the laboratory by Lyman and Paschen.

A philosophical lesson to draw from this is as follows. It's possible to spot surprisingly complex empirical laws (or regularities)—laws concerning observable things, such as the gaps between lines in images produced by relatively complicated experiments—without having a means to derive said laws from posits about unobservable things. Indeed, it's possible to spot such laws without even having a sophisticated story to tell about the unobservable components of the observable things governed by the law. Moreover, doing so is progressive in its own right.

Finding new empirical laws may be progressive as a 'stepping stone' too, say for finding ways to connect several existing empirical laws with appropriate narratives about unobservable things, or generating 'umbrella' laws involving unobservable posits. The findings of Balmer and Rydberg proved useful in this respect, as we'll see later.

2. An Abundance of Atoms: Thomson, Nagaoka, Kelvin, Nicholson, . . .

We now come back to the time of the Zeeman effect. After the contemporaneous discovery of the electron—which 'swept through physics almost as fast as it ran down wires' (Aaserud and Heilbron 2013: 117)—models of the atom proliferated.[5] Thomson's inappropriately named 'plum pudding' model—on which 'the atom consists of a number of corpuscles moving about in a sphere of uniform positive electrification . . . in a series of concentric shells' (Thomson 1904: 255)—and Nagaoka's Saturnian model—which 'differs from the Saturnian system considered by Maxwell in having repelling particles instead of attracting satellites' (Nagaoka 1904: 445)—are the best known. But Kelvin (1902) offered a different model (which *was* like a plum pudding), and more promising alternatives, such as Nicholson's (1911) 'ring' model, appeared subsequently.[6] Some such models, which later appeared to be on the right track, were discarded on the basis of classical theoretical considerations. For example, Nagaoka's atom was mechanically unstable—Heilbron (1977: 53) rather harshly, but not inaccurately, declares that Nagaoka 'blundered in adapting Maxwell's prize-winning investigation of the stability of Saturn's rings to the electromagnetic case'—in virtue of the *repulsive* forces between the satellites. The clue that it might be on the right track, to some extent, was to come from new experimental findings.

I'll come to those experimental findings in the next section. But before I do, I should like to emphasise two things. First, most of the aforementioned atomic models, just like the one we'll later consider in further depth, namely Bohr's, were semi-literal—or even non-literal—in character. In Thomson's atomic model, for instance, the positive charge was a non-literal component: 'the negative effect is balanced by something which causes the space through which the corpuscles are spread to *act as if* it had a charge of positive electricity' [emphasis added] (Thomson 1899: 565). Or as Kragh (2012: 16) puts it: 'The imponderable sphere of positive electricity was a mathematical artifice, a ghost-like entity'.

More generally, in the words of Heilbron (1977: 41–43):

> [T]he representations were not meant or taken literally. . . . The same physicist might on different occasions use different and even conflicting pictures of the same phenomena . . . piecemeal analogies or provisional illustrative models.

And this echoes the words of Poincaré nearer the time, in his *Électricité et Optique* (quoted by Duhem 1954: 85):

> The English scientist does not seek to construct a single, definitive, and well-ordered structure; he seems rather to raise a great number of provisional and independent houses among which communication is difficult and at times impossible.

Such actual practice fits nicely with the account of scientific progress given in the first chapter—that is, if what was taken to be progressive by the British scientists, at that time, was typically so. The modelling process didn't involve a strong push for consistency, in so far as there was no demand for a true—or even a merely approximately true—account of the unobservable.

Second, in harmony with the account in the first and second chapters, these Cambridge-influenced Victorian scientists emphasised the importance of such a modelling approach for fostering *understanding*.[7] Kelvin (1884: 270), for example, declared:

> If I can make a mechanical model, I understand it. As long as I cannot make a model all the way through I cannot understand. . . . I want to understand light as well as I can without introducing things that we understand even less of.

And Lodge (1892: 13) explicitly connected such thoughts about models to the role of analogies:

> [I]f we resist the help of an analogy . . . there are only two courses open to us: either we must become first-rate mathematicians, able to live wholly among symbols, dispensing with pictorial images and such adventitious aid; or we must remain in hazy ignorance of the stages which have been reached, and of the present knowledge.

The analogies (and hence models) employed were, however, limited in one significant respect. They were, as emphasised by Kelvin, mechanical. Heilbron (1977: 41) makes this limitation more precise:

> The single effective constraint was that only those classes of forces with which physicists had become familiar since the time of Newton should be admitted; and it went without saying that the resultant description had to be continuous in space and time.

This constraint may profitably be discarded; it was violated by Bohr's discontinuous model of the atom, as we will see. However, as Bailer-Jones (2009: 41–42) points out:

> [T]here are contemporary conceptions of mechanism that preserve what is attractive about mechanical models and what it is about them that may make us understand. These conceptions involve the abstraction of the elements of the classical nineteenth-century mechanism. . . . In the light of such revised conceptions, it becomes possible to highlight the benefits of mechanical models for understanding without being retrograde about one's understanding of modern science.

My own way of approaching this, as discussed in Chapter 2, is to focus on the role of familiar observable things (and observable processes involving these) in the modelling process, rather than 'mechanisms' in any deeper sense.

So to wrap up this section, the way that much of Victorian science was conducted gels well with the theses advanced in the previous chapters, and especially the first and the second. It gels in the precise sense that the science of the period counts as good science—conducted, moreover, by scientists who'd have been inclined to agree with much of what I've said about progress and models—on a cognitive instrumentalist view. Conflicting models were used simultaneously, without any push for consistency. Such models were used with the explicit aims of furnishing understanding (rather than explanation in any deeper, factive or quasi-factive, sense) and suggesting new research avenues. Furthermore, they typically involved analogies with observable things (especially mechanical things). In summary:

> [T]he principal English pedagogues of physics considered a theory incomplete without an accompanying model or analogy, ideally elaborated to the last detail. Such pictures, they believed, fixed ideas, trained the imagination, and suggested further applications of the theory. . . . This pedagogical bias was built into the training of British physicists.
>
> (Heilbron 1977: 42)

3. Microscopes in the Dark: Geiger, Marsden, and Rutherford

Next came an important series of experiments by Geiger and Marsden, conducted under the guidance of Rutherford, an ex-student of Thomson's, between 1908 and 1913.[8] The lion's share of the credit should go to the youngsters. Rutherford did not want to spend long periods of time in a dark room, looking through a microscope for minute scintillations on a fluorescent screen. He remarked in 1908, in a letter to Bumstead:

> Geiger is a good man and worked like a slave. I could never have found time for the drudgery before we got things going in good style. . . . Geiger is a demon at the work of counting scintillations and could count at intervals for a whole night without disturbing his equanimity. I damned vigorously and retired after two minutes.
>
> (Eve 1939: 180)

In the most remarkable of the experiments, which occurred in 1909, alpha particles—which Rutherford had recently found to be helium nuclei—were fired at wafer-thin metal foil and deflected by over 90 degrees on some occasions. The occasions were rare. They were around 1 in 8,000 in the case of platinum. The experiment must have been especially tiresome to conduct. Rutherford did well to avoid doing it. Remarkably, it is still often known as 'the Rutherford experiment'.[9]

The actual authors of the key paper concluded:

> If the high velocity and mass of the α-particle be taken into account, it seems surprising that some of the α-particles, as the experiment shows, can be turned within a layer of 6 x 10^{-5} cm. of gold through an angle of 90°, and even more. To produce a similar effect by a magnetic field, the enormous field of 10^9 absolute units would be required.
>
> (Geiger and Marsden 1909: 498)

Rutherford's (1938: 68) analogical 'recollection' is more dramatic:

> It was quite the most incredible event that has ever happened to me in my life. It was almost as incredible as if you fired a 15-inch shell at a piece of tissue paper and it came back and hit you. On consideration, I realized that this scattering backward must be the result of a single collision, and when I made calculations I saw that it was impossible to get anything of that order of magnitude unless you took a system in which the greater part of the mass of the atom was concentrated in a minute nucleus. It was then that I had the idea of an atom with a minute massive centre, carrying a charge.

However, this really is a 'recollection', rather than a recollection, as explained by Heilbron (1968: 265):

> That the military imagery and the incredulity are later fabrications we can easily see from a lecture Rutherford delivered in September, 1909, six months after the discovery of the diffuse reflection . . . [in which he said]:
>
>> Geiger and Marsden observed the surprising fact that about one in eight thousand α particles incident on a heavy metal like gold is so deflected by its encounters with the molecules that it emerges again on the side of incidence. Such a result brings to light the enormous intensity of the electric field surrounding or within the atom.
>
> The reflection seems not yet to have become the most remarkable event in Rutherford's life![10]

So Rutherford did not swiftly conclude that there was a 'minute massive centre'. And even when he had settled on the view that there was such a centre, Rutherford did not assume that it must be positively charged, as is often claimed. On the contrary, he recognised the possibility of a 'slingshot' effect due to a concentration of negative charge; for as Andrade

(1962: 38) explains: 'the angle through which an alpha particle, aimed to pass near the charged nucleus . . . is deflected, is independent of the sign of the charge'. In 1911, Rutherford even wrote to Bragg: 'I am beginning to think that the central core is negatively charged, for otherwise the law of absorption for beta rays would be very different from that observed' (Eve 1939: 194–195). The ascription of positive charge to the nucleus, around 1912, was based on additional data.[11] The terrain was rough. The ground was boggy. Visibility was poor. There were no signs pointing towards Truth, or even towards Empirical Adequacy.[12]

Analogies between the atom and the solar system had been made several times, independently, much earlier. This is as we'd expect, given the importance of analogies involving observables in science, and the structural similarity between the equations for gravitational and electrostatic force. As Kragh (2012: 3) explains, 'Colding . . . [i]n an unpublished note of 1854 . . . pictured atoms, or what he called "molecules", as analogous to planetary systems.'[13] Slightly later, Weber developed another planetary model. And he got several things right, from the perspective of contemporary theory, e.g. that equal and opposite charges, but different characteristic masses, are possessed by (some) atomic constituents. As early as 1871, for instance, he wrote:

> Let e be the positively charged electrical particle, and let the negative particle, carrying an opposite charge of equal amount, be denoted $-e$. Let only the latter be associated with the massive atom, whose mass is so large that the mass of the positive particle may be considered negligible. The particle $-e$ may then be considered as being at rest, while just the particle e moves around the particle $-e$.
> (Weber 1871: 44, translated by Mehra and Rechenberg 1982: 169)[14]

The findings of Geiger and Marsden placed the planetary analogy on a more solid footing. That is to say, in so far as the results of their experiments could be understood to be a consequence of the interactions of alpha particles with a nucleus. Crucially, this wasn't just a qualitative matter. Rutherford's planetary model predicted the *frequency* of deflections greater than 90 degrees to be expected.[15] Hence, the planetary view of the atom did work that it had not when it was earlier proposed.

4. The Great Dane

> One of Bohr's great strengths as a theorist was the ability, which he shared with Thomson, to imagine himself as a participant in the models he analyzed . . . an exceptional capacity to share the experiences of the microworld.
>
> —Aaserud and Heilbron (2013: 133)

The stage was set for Bohr's (1913) model of the hydrogen atom, and derivative models of the atoms of heavier elements. (I haven't given the whole story. An emerging consensus at the first Solvay conference about the limitations of classical physics was significant, as was Bohr's connection with Rutherford.[16] The influence of Nicholson's work is also noteworthy.[17]) The core was a positively charged 'nucleus'. But what of the electrons? They couldn't be static, because the atoms would be *mechanically* unstable due to the repulsive forces between the electrons; this result, due to Earnshaw (1842: 106), was well known: 'instability cannot be removed by *arrangement* . . . there will always exist a direction of instability'.[18] But if they accelerated, and were relatively few in number, then the atoms would be *electromagnetically* unstable due to the resultant energy loss, in line with Maxwell's finding that accelerating charges emit radiation. What's more, one would expect the emission spectra to be continuous on account of the changing acceleration as the orbit degraded.[19] (If there were enough electrons, the energy drain could be sufficiently limited due to shielding. Yet as mentioned earlier, *mechanical* instability due to the repulsive forces between the electrons would result nonetheless. Recall Nagaoka's 'blunder'.)

Bohr's way around these difficulties was to *assume* stability, and to see what follows if energy emissions are quantised. Moreover, he assumed that the stable situation could be described, at least to some extent, by classical mechanics. So he employed a semi-classical approach, based, in his own words, on the posits:

(1) That the dynamical equilibrium of the systems in the stationary states can be discussed by help of the ordinary mechanics, while the passing of the systems between different stationary states cannot be treated on that basis.

(2) That the latter process is followed by the emission of a *homogeneous* radiation, for which the relation between the frequency and the amount of energy emitted is the one given by Planck's theory.

(Bohr 1913: 7)

The model is a departure from the typical Victorian style of modelling in so far as it's discontinuous in character. That's to say, it (explicitly) involves no mechanical account of how 'the systems [pass] between different stationary states'.[20] And this is a downside, from the point of view of promoting understanding. As Rutherford put it, in a letter to Bohr:

The mixture of Planck's ideas with the old mechanics makes it very difficult to form a physical idea [that is, picture] of the basis of it all. There appears to be one grave difficulty in your hypothesis, which I have no doubt you fully realize, namely, how does an electron decide what frequency it is going to vibrate at . . . when it passes from one stationary state to the other? It seems to me that you would have to assume that the electron knows beforehand when it is going to stop.

(Quoted—with the additional note about 'physical idea' meaning 'picture'—in Heilbron 2003: 73)

The upside, however, was that the model could be employed not only to connect atomic theory with previous experimental spectroscopic findings in a remarkably accurate fashion, but also to predict new findings. It compromised on understanding and external consistency—that is, consistency with some established theories of the day—but was worthwhile precisely because of these rewards. As Poincaré (quoted by Aaserud and Heilbron 2013: 146) put it, theories (and hence models) 'must establish a connection among the various experimental facts and, above all, support predictions.'[21] Admiring this model is consistent with thinking that promoting understanding is important, and that ideally the model would have been easier to picture. Recall my argument that good modelling is about finding sweet spots, in Chapter 1. This almost invariably involves compromises. And sometimes a spot is sweet enough for some purposes, but not for others. This is why multiple models for the same system may be useful.

Now there's considerable historical evidence that Bohr was strongly influenced by the Victorian style of modelling. Beyond the relatively circumstantial—Bohr's admiration for Thomson, and failed attempts to secure his support—Aaserud and Heilbron (2013: 114) note, for instance, that:

> The awkward advice to use the concepts of mechanics as illustrative or analogic where not fully competent suggests the method of the Trilogy. Bohr studied *Aether and Matter* carefully during his second term at Cambridge and ranked it very highly. "When I read something that is so good and grand [as Larmor's book], then I feel such courage and desire to try whether I too could accomplish a tiny bit."
> [Quotation in speech marks from Bohr's correspondence]

Larmor not only used the Victorian style—like any good Cambridge physicist of the time[22]—but also championed it in *Aether and Matter*, which appeared in 1900. Here, for instance, is a passage from the preface:

> The accumulation of experimental data, pointing more or less exactly to physical relations of which no sufficiently precise theoretical account has yet been forthcoming, is doubtless largely responsible for the prevalent doctrine that theoretical development is of value only as an auxiliary to experiment, and cannot be trusted to effect anything constructive from its own resources except of the

kind that can be tested by immediate experimental means. The mind will however not readily give up the attempt to apprehend the exact formal character of the latent connexions between different physical agencies: and the history of discovery may be held perhaps to supply the strongest reason for estimating effort towards clearness of thought as of not less important in its own sphere than exploration of phenomena.

(Larmor 1900: ix)

There are other remarkable forays into epistemology in the fifth chapter—'On method in general physical theory'—such as:

It is not superfluous to consider sometimes what there is to prevent many of the scientific hypotheses . . . of Natural philosophy, which are at present effective and successful, from being . . . of a merely provisional and analogical character. The uniformities which it is customary to call laws of nature are often just as much laws of mind: they form an expression of the implications between mind and matter, by means of which material phenomena are mentally grasped. . . . The formal analogies between the mathematical theories of different branches of physics perhaps originate as much in the nature of the necessary processes of thought as in the nature of the external things: for 'the mind sees in all things that which it brings with it the faculty of seeing.'

(Larmor 1900: 70)[23]

It's reasonable to think that passages such as this influenced Bohr somewhat. But let's now look at some of his admirable results. I won't stick closely to any of Bohr's own derivations, because my aim is to make the illustration as accessible as possible (by starting from relatively elementary foundations).[24] Purely for illustrative purposes, I'll also use reasoning that wasn't available to Bohr at one point. However, I'll make it explicit when I do this, and show how the same result could be achieved in historical context, albeit in a less elegant fashion, in an addendum to this chapter.

Consider a single electron in a circular orbit around a nucleus with equal and opposite charge. Assume the orbit is stable. (Also ignore gravitational forces, and so forth. Such idealisations are hardly ever mentioned, which reinforces how commonplace they are in physics. Physicists expect them, and acquire a sense of which to buy into on a context-by-context basis.) Let k represent Coulomb's constant (i.e., $1/4\pi\varepsilon_0$), e represent the charge on the electron, and r represent the orbital radius. The potential energy of the electron, U, is then:

$$U = -\frac{ke^2}{r} \tag{1}$$

And the kinetic energy of the electron, K, is simply a function of its mass, m, and its velocity, v:

$$K = \frac{mv^2}{2} \tag{2}$$

Hence the total energy of the electron, E, is as follows:

$$E = K + U = \frac{mv^2}{2} - \frac{ke^2}{r} \tag{3}$$

Now let's consider the forces in play. Mechanically, the centripetal force on the electron is:

$$F = \frac{mv^2}{r} \tag{4}$$

And electrostatically, according to Coulomb's law, the force on the electron is:

$$F = \frac{ke^2}{r^2} \tag{5}$$

It follows from (4) and (5), by substitution, that:

$$\frac{ke^2}{r^2} = \frac{mv^2}{r} \tag{6}$$

And we may rewrite (6) as:

$$\frac{ke^2}{r} = mv^2 \tag{7}$$

We can now consider the energy equations alongside the force equations. We see, by substitution of (7) into (1), that

$$U = -mv^2 \tag{8}$$

Indeed, it was a well-known result that the potential energy is double the kinetic energy, for bodies in circular motion in analogous gravitational contexts. And it follows from (8) and (3) that:

$$E = -\frac{mv^2}{2} \tag{9}$$

All the reasoning so far was as trivial for any early twentieth-century physicist as it is now. But we now come to the more interesting bit. Angular momentum, L, is defined as follows:

$$L = mvr \tag{10}$$

Thus, by substitution from (7), we can see that:

$$ke^2 = Lv \tag{11}$$

And we may rewrite this equation as follows:

$$v = \frac{ke^2}{L} \tag{12}$$

Thus, we can see that the total energy in terms of angular momentum—by substitution from (12) into (9)—is:

$$E = -\frac{mk^2e^4}{2L^2} \tag{13}$$

It follows that the energy difference between two such stable states, where the angular momenta of the electrons are L_1 and L_2, and the former is higher than the latter, is:

$$\Delta E = \frac{mk^2e^4}{2}\left[\frac{1}{L_1^2} - \frac{1}{L_2^2}\right] \tag{14}$$

It's plain to see that the right hand side of this equation has a structural similarity with the right hand side of (0), namely Rydberg's equation for spectral lines. And if we assume that this energy difference corresponds to the emission energy, we know that:

$$E = hf = \frac{hc}{\lambda} \tag{15}$$

I'll now use an idea that Bohr didn't have at his disposal, which was presented in De Broglie's PhD thesis of 1924. I might instead have misled you by saying that Bohr assumed that electron angular momentum is quantised. But he didn't, although this is often claimed in contemporary physics textbooks (and other places as a result). Rather, Bohr assumed that electron orbits were stable, and derived the fact that angular momentum is quantised. This feature of his reasoning process is faithfully represented by the derivation that follows, involving equations (16) to (19).

It's also represented by the more lengthy derivation in the addendum to this chapter, which has the virtue of avoiding appeal to any theoretical notions that weren't available to Bohr in 1912.

The idea we'll use is matter waves, or that matter has wave-like properties. According to De Broglie (1925), for an electron, like any other massive particle,

$$p = \frac{h}{\lambda} \tag{16}$$

Hence, given that p classically represents mv, it follows, from (10) and (16), that:

$$L = \frac{rh}{\lambda} \tag{17}$$

Now in order for an electron's orbit about a nucleus to be stable, it must form a *standing wave*. That's to say, the number of wavelengths over the whole length of the wave must be equal to a positive integer. The whole length of the wave is simply the circumference of the orbit, $2\pi r$. De Broglie (1925: 28–29) put it like this:

> It is physically obvious, that to have a stable regime, the length of the channel must be resonant with the wave; in other words, the points of a wave located at whole multiples of the wave length, l, must be in phase. The resonance condition is $l = n\lambda$ if the wavelength is constant. . . . [T]he resonance condition can be identified with the stability condition from quantum theory. . . . From this we see why certain orbits are stable.

In the present notation, that's to say,

$$\lambda = \frac{2\pi r}{n} \tag{18}$$

Thus, by substitution from (18) and (19),

$$L = \frac{nrh}{2\pi r} = \frac{nh}{2\pi} \tag{19}$$

All changes between stable states must therefore involve changes in angular momentum of $h/2\pi$, or multiples thereof.

We almost have our final result. Consider again an electron moving between two stable states, in which its angular momentum is L_1 and L_2 respectively (and the former is higher than the latter). Let the number of wavelengths in each state be represented by n_1 and n_2. Now we may substitute into (14), from (19), as follows:

$$\Delta E = \frac{mk^2e^4}{2}\left[\frac{1}{\left(\frac{n_1b}{2\pi}\right)^2} - \frac{1}{\left(\frac{n_2b}{2\pi}\right)^2}\right]$$

$$= \frac{mk^2e^4}{2}\left[\frac{4\pi^2}{n_1^2b^2} - \frac{4\pi^2}{n_2^2b^2}\right]$$

$$= \frac{2\pi^2mk^2e^4}{b^2}\left[\frac{1}{n_1^2} - \frac{1}{n_2^2}\right] \tag{20}$$

Now look back to (15), and note that any change of energy (ΔE) will, of course, be an energy value. Hence, in our penultimate step, we can rewrite (20) as:

$$\frac{1}{\lambda} = \frac{2\pi^2mk^2e^4}{b^3c}\left[\frac{1}{n_1^2} - \frac{1}{n_2^2}\right] \tag{21}$$

So it is evident, with reference back to the Rydberg equation (0)—given, recall, that k represents $1/4\pi\varepsilon_0$—that we have a 'reductive' account of the Rydberg constant:

$$R = \frac{2\pi^2mk^2e^4}{b^3c} = \frac{me^4}{8\varepsilon_0^2b^3c} \tag{22}$$

It would be hard to overestimate the beauty and potency of this result in historical context. It connects a remarkable array of repeatable experiments (in so far as it holds). That's because R may be measured independently of any factor on the right-hand side of the equation, and various clusters of factors on the right-hand side may be measured independently of R and each other. It's a triumph that the numbers add up (if indeed they do). The force of this equation was also predictive, and not merely accommodative (or 'explanatory'); new ways to measure the factors in the equation became available. With a slight addition—introduction of a factor, Z, to represent the number of positive unit charges in the nucleus—it can also be extended to 'hydrogenic' atoms or ions (i.e., cases where electron shielding is not significant), such as He+ (for which Z is two). Indeed, Bohr was able to use his model to show how lines from the ζ Puppis star, which were taken to be due to hydrogen by Pickering (1897) and Fowler (1912), were in fact from ionised helium.[25] Kragh (2012: 68–74) explains this episode in considerable detail.

5. The Advent of Spin

> A colleague who met me strolling rather aimlessly in the beautiful streets of Copenhagen said to me in a friendly manner, "You look very unhappy," whereupon I answered fiercely, "How can one look happy when he is thinking of the anomalous Zeeman effect?"
>
> —Pauli (1946: 214)

Despite being a great success in several respects, Bohr's model still couldn't satisfactorily account for some phenomena, such as the Zeeman effect with which we began. The Stark (1913) effect, an analogous effect caused by electric rather than magnetic fields, was also an important stimulus for subsequent developments. The first steps in improving the model were taken by Sommerfeld and Debye in the next couple of years.[26] They introduced two extra 'quantum numbers', in addition to n. To put it simply, Bohr's model was extended so that it covered elliptical orbits. One of the new quantum numbers related to the shape of the orbits. The other related to the orientation of the orbits. These posits weren't to be taken literally, according to Debye, who remarked (as reported by Gerlach): 'surely you don't believe that the orientation of the atoms is something real; that is a prescription of the calculation—a timetable of the electrons' (Mehra and Rechenberg 1982: 435).[27]

A little later, in 1920, Sommerfeld introduced a fourth quantum number, an 'inner' quantum number, which he took to be related to the angular momentum of the atom (in a way we'll see below). However, the so-called *anomalous* Zeeman effect, exhibited by elements such as sodium, still wasn't satisfactorily accounted for. Partly as a result, alternative models were developed by Landé and Pauli in the early 1920s, as detailed by Tomonaga (1997: ch. 1); each had different quantum number systems, with advantages and disadvantages relative to the others.[28] We can pass over the fine details of these, although the episode illustrates how difficult model making can be. We need only note that there was disagreement, in particular, about what the quantum numbers should be understood to physically represent.

One key idea, championed by Landé, was that the Zeeman effect might be explained by an interaction between the core and the excited electron. (We'll only consider cases where a single electron is excited, for simplicity's sake.) He therefore took the inner quantum number to depend on the core's angular momentum. However, this model had some odd consequences.

Consider a sodium atom (Na) and a singly ionised magnesium cation (Mg^+). Each has the same number of electrons, and, thus, as one might expect on the model, a similar spectrum to the other. One would therefore expect that the core of the magnesium atom would have the same angular momentum as the cation. Yet it does not. Rather, as Tomonaga (1997: 20) notes: '[I]f an electron approaches Mg^+ from the outside and the ion becomes a core of Mg, suddenly *J* [the inner quantum number] must change from 1 to either 1/2 or 3/2.' Or to put matters less technically (Tomonaga 1997: 30): 'if you adopt the core concept, some property of the core must suddenly change when an electron is added to the core.'

Tomonaga (1997: 30) calls this consequence an 'absurdity', and notes that it played a role in Pauli's rejection of the model, although it didn't especially trouble Landé. I sympathise with Landé, because a physicist of the time, following in the footsteps of Bohr and Sommerfeld, might reasonably have thought as follows.[29] We have made a move to semi-classical modelling in atomic physics, and this move has led to unprecedented predictive successes in the domain. Moreover, a semi-classical model may involve specifying that changes occur in some circumstances without providing a physical mechanism by which those changes occur. Transitions in energy levels, in Bohr's model, are like this. (And ultimately, as we will see, the weirdness just gets shifted elsewhere. The weirdness was not even minimised in the Copenhagen view that later became dominant. As Chapter 2 explains, Bohm's version of quantum mechanics is less weird.)

The alternative view developed by Pauli was that the third and fourth quantum numbers are related to properties of the excited electron:

> The closed electron configurations shall not contribute to the magnetic moment and the angular momentum of the atom. Especially in the case of the alkalis [i.e., alkali metals], the value of the angular momentum of the atom and its change in energy in an external magnetic field are considered to be mainly due to the radiant [or excited] electron, which is also thought to be the origin of the magneto-mechanic anomaly. From this point of view the doublet structure of the alkali spectra . . . is caused by a strange two-valuedness of the quantum-theoretical properties of the radiant [or excited] electron which cannot be described classically.
>
> (Translated from the original and quoted in Tomonaga 1997: 29–30)

This move prepared the way for the idea that electrons might 'spin', or self-rotate, which was proposed by Kronig after reading one of Pauli's letters to Landé.[30] He was discouraged from publishing the idea, however, by a subsequent personal encounter with Pauli (who had a well-deserved reputation for being a harsh critic).[31] Kronig must have been irritated when Uhlenbeck and Goudsmit (1925) proposed electron spin—a possibility that they conceived of independently—only a year later. (As it happens, Uhlenbeck tried to withdraw this paper after submitting it, due to negative comments received from Lorentz. But he was too late.) He could take some small pleasure, nonetheless, in explaining some of the problems with the idea; see Kronig (1926) and Mehra and Rechenberg (1982: 692–703) for more on this story.

One significant problem was accuracy; key predictions about spectral intervals were out by a factor of two, as explained by Uhlenbeck and Goudsmit (1926). However, Thomas (1926) solved this problem by showing that the calculations involved in generating those predictions were faulty, because they weren't conducted in the correct reference frame.[32] So let's put this problem to one side. The most obvious remaining objection is that an electron would have to be spinning much faster than is possible, according to special relativity (SR) and standard estimates of its size, in order to have such a high intrinsic angular momentum. In fact, its surface would have to be moving 'ten times higher than that of light' (Tomonaga 1997: 35).[33] Thus one couldn't take a story involving all of the aforementioned components literally; and since the physicists of the day were inclined to take SR and the size estimates as correct, they were inclined not take the talk of spin literally. In short, if electron spin of *that* kind had been detected—or even observed (albeit not in the philosopher's sense)—then relativity had plausibly been falsified. *But Thomas precession is a special relativistic effect.* Thus the predictive success of the spin posit *depended* on special relativity in a non-trivial sense; there was a genuine tension in trying to renounce the latter while accepting the former. (There was no conceived alternative to special relativity that would do the trick of explaining the factor of two. We can also note that the spin-statistics theorem was later derived on a special relativistic basis; so holding on to special relativity proved to be fruitful.)

I'll now summarise some of the key findings of this section, and draw a few lessons from them. (For simplicity's sake, in line with the previous discussion, I'll just write of electrons; that is, although spin quantum numbers are assigned to other theoretical entities too.) First, the idea of spin (*qua* self-rotation) was prompted by Pauli's idea that the third and fourth quantum numbers were closely related to, or represented, electron properties. More particularly, it was encouraged by the notion that the fourth quantum number might represent an intrinsic property of the electron. Second, the idea of spin (*qua* self-rotation) proved useful in making predictive progress, and especially in working out the proper (*qua* predictively successful) values for the fourth quantum number. Third, however, the existence of a self-rotating electron was in conflict with a theory, namely special relativity, responsible for the predictive progress. More particularly, using special relativity to think about how the electron would move relative to the atom led to a nice result (i.e., explained away a problematic factor of two). Alas, thinking that the electron *really* rotated was inconsistent with this very theory (which entails that nothing can move as fast as, let alone faster than, light).[34]

In line with the discussion of spin in Chapter 2, the talk of the electron having an unobserved intrinsic property—of a categorical variety—*may* be taken literally. And one could reasonably think of spin in such a way, and then associate various dispositions of the electron, described in terms of observable properties, with this intrinsic property. The notion that spin is an intrinsic angular momentum—as opposed to an intrinsic angular momentum *number*, as Pauli would have had it at some points in his career—is a different matter, because of the way that the concept of intrinsic angular momentum is related to the concept of movement about a point. Quite literally, it involves a *moment*.

If one insists on taking 'spin' to refer to an (unobserved) intrinsic property of the electron, however, then one may reasonably believe—according to findings on Bohm's theory in Chapter 2, and the issues concerning unconceived alternatives covered in Chapter 3—that spin does not exist. That is, although the fourth quantum number is an important predictive tool (as is the later spin operator) and talk of an intrinsic property of the electron is a useful device. Recall how, in Bohm's theory, the position of an electron (or other entity with spin) and the appropriate wavefunction are enough to explain its behaviour, say in Stern-Gerlach experiments:

> The particle itself, depending upon its initial position, ends up in one of the packets moving in one of the directions.
>   The probability distribution . . . can be conveniently expressed in terms of the quantum mechanical spin operators. . . . From a Bohmian perspective there is no hint of paradox in any of this—unless we assume that the spin operators correspond to genuine properties of the particles.
>
> (Goldstein 2013)

Incidentally, the history of the Stern-Gerlach experiment is interesting with respect to some of the findings in Chapter 3. The experiment was performed with the aim of confirming the Bohr-Sommerfeld model of the atom, and was taken to confirm it, although it did just the opposite. Friedrich and Herschbach (2003: 57) write:

> The gratifying agreement of the Stern-Gerlach splitting with the old theory proved to be a lucky coincidence. The orbital angular momentum of the silver atom is actually zero, not $h/2\pi$ as presumed in the Bohr Model. . . . The magnetic moment is nonetheless very nearly one Bohr magneton, by virtue of the Thomas factor of two, not recognized until 1926. Nature was thus duplicitous in an uncanny way.
>   A curious historical puzzle remains. In view of the interest aroused by the SGE [Stern-Gerlach experiment] in 1922, we would expect that the postulation of electron spin in 1925 should very soon have led to a reinterpretation of the SGE splitting as really due to spin. However, the earliest attribution of the splitting to spin that we have found did not appear until 1927, when Ronald Fraser noted that the group-state orbital angular momentum and associated magnetic moments of silver, hydrogen, and sodium are zero. Practically all current textbooks describe the Stern-Gerlach splitting as demonstrating

electron spin, without pointing out that the intrepid experimenters had no idea it was spin that they had discovered.

Science is extremely difficult, despite what we were taught at secondary school (and even during science degrees). What an experiment suggests—not 'shows'—is theory-laden, as emphasised in Chapter 3.

6. Conclusion

We have seen how a significant period in the history of science bears on several of the views defended in previous chapters.

First, we saw that many leading scientists in this period adopted views on scientific progress—in particular, on the central importance of predictive power and understanding—which are consistent with key elements of the account of scientific progress advocated in Chapter 1. This casts doubt on the claim that scientific realism is, and especially that it was, 'the sciences' own philosophy of science', as Boyd (2002) claims 'most of its defenders' think.[35]

Second, we saw how predictive progress doesn't always depend on theoretical (or technological) progress, and particularly on posits about unobservable things. Consider again the empirical laws of Balmer and Rydberg, concerning the spectrum of hydrogen.

Third, we saw how it is possible to gain an understanding of empirical laws without generating a true, or even an approximately true, model. That is, on the assumption that Bohr's model isn't approximately true of hydrogen, as it presumably can't be if the contemporary view of the atom is approximately true (as realists tend to think it is).[36] Despite being false, and having many intentional gaps, the model put some theoretical flesh on the Rydberg constant, and had remarkable unificatory power (especially at the level of connecting different phenomena).

Fourth, and finally, we've seen how developments in science may be fraught with difficulty, and how there's much more scope for reasonably interpreting experimental results than textbook histories would lead us to believe. This came across, for instance, in the consideration of the famous 'Rutherford experiment'—the one performed by Geiger and Marsden. Settling on a sign for the charge on the nucleus—even settling on there being a nucleus—was no easy matter.

Addendum — A Derivation of the Quantisation of Electronic Angular Momentum[37]

I'll begin by reproducing several equations introduced previously, which we'll require again in the following.

$$\frac{ke^2}{r} = mv^2 \tag{7}$$

$$L = mvr \tag{10}$$

$$E = -\frac{mk^2e^4}{2L^2} \tag{13}$$

$$E = hf \tag{15}$$

Consider an electron in an extremely high energy level in the hydrogen atom. The predictions for the emission frequency in a quantised model—or any empirically adequate model—should approximate to the classical predictions. Thus, the emission frequency should be approximately the same as the orbital frequency.

Classically, for a particle in circular motion, the velocity, v, is related to the angular velocity, ω, as follows:

$$v = r\omega \tag{23}$$

Moreover, the angular velocity is related to orbital (and hence emission) frequency, f, like so:

$$\omega = 2\pi f \tag{24}$$

Thus, by substitution of (24) into (23), we can express the orbital (and hence emission) frequency of the particle as follows:

$$f = \frac{v}{2\pi r} \tag{25}$$

Hence, by substitution from (25) into (15):

$$\Delta E = \frac{hv}{2\pi r} \tag{26}$$

Now the energy differences between levels (or allowable *qua* stable states) for an electron such as one we're considering will also be very close, and may therefore be represented by a differential. So from (13):

$$\Delta E \cong \frac{dE}{dL}\Delta L = \frac{mk^2e^4}{L^3}\Delta L \tag{27}$$

And from this, by substitution from (26), it follows that:

$$\frac{hv}{2\pi r} \cong \frac{mk^2e^4}{L^3}\Delta L \tag{28}$$

We may now replace L in (28), in accord with (10):

$$\frac{h\upsilon}{2\pi r} \cong \frac{mk^2e^4}{m^3\upsilon^3r^3}\Delta L = \frac{k^2e^4}{m^2\upsilon^3r^3}\Delta L \tag{29}$$

We are almost done. By rearranging (7), we arrive at:

$$1 = \frac{m\upsilon^2r}{ke^2} = \frac{m^2\upsilon^4r^2}{k^2e^4} \tag{30}$$

Thus, by substitution from (30) into (29)—and a cheeky drop of '$\cong$'[38]—we find:

$$\frac{h}{2\pi} = \Delta L \tag{31}$$

Notes

1 So I agree with Arabatzis (2001: 173), writing on the 'discovery' of the electron, that:

> to talk about the discovery of an unobservable entity one has to face a difficulty that does not appear in the case of observables. The discovery of an observable entity might simply involve its direct observation and does not require that all, or even most, of the discoverer's beliefs about it are true. . . . This is not the case, however, when it comes to unobservable entities where direct physical access is, in principle, unattainable. The lack of independent access to such an entity makes problematic the claim that the discoverer's beliefs about it need not be true. If most, or even some, of those beliefs are not true it is not evident that the "discovered" entity is the same with [sic] its contemporary counterpart. It has to be shown, for instance, that Thomson's "corpuscles," which were conceived as classical particles and structures in the ether, can be identified with contemporary "electrons," which are endowed with quantum numbers, wave-particle duality, indeterminate position-momentum, etc. This would require, among other things, a philosophical theory of the meaning of scientific terms that would enable one to establish the referential stability of a term, despite a change of its meaning.

This explains why the causal theory of reference is explored by Psillos (1999) and other realists. For an opposing view to Arabatzis's, and a putative account of discovery, see Achinstein (2001). See also Falconer (1987) and Arabatzis (2006: ch. 3).

2 Any such point at which to begin is somewhat arbitrary, although I am forced to decide on such a point in the interests of brevity. Spectroscopy began to develop beyond its Newtonian infancy considerably earlier, with the work of Melvill, Wollaston, and especially Fraunhofer. Yet it is fair to say that it was adolescent, at best, until the 1850s. As Dingle (1963: 199) puts it: '[in the 1850s] the science of spectroscopy, though not yet christened, may be said to have attained its majority and to be just entering on its period of full adult development.' Of special importance, after this point, were two

changes. First was the recognition of a significant relationship between emission and absorption spectra, due Ångström and Kirchoff. The former, writing in 1855, suggested that: 'the explanation of the dark lines in the solar spectrum contains the explanation of the bright lines in the electrical' (Reif-Acherman 2014: 17). Second was the increased accuracy and precision in measurements. In the case of the hydrogen atom, as we will see, Ångström's measurements of the wavelengths of the first four visible lines, in 1871, were to prove especially important.

3 In the words of Balmer (1885: 81–83):

> From the formula we obtained for a fifth hydrogen line 49/45.3645.6 = 3969.65.10^{-7} mm. I knew nothing of such a fifth line . . . and I had to assume that either the temperature relations were not favourable for the emission of this line or that the formula was not generally applicable. On communicating this to Professor Hagenbach he informed me that many more hydrogen lines are known, which have been measured by Vogel and by Huggins.

4 A derivative formula holds for atoms or ions that are 'hydrogenic' (or suitably similar to hydrogen). Key examples are ions with only one electron, such as He^+ and $Li2^+$.

5 In the view of Heilbron (1977: 40): 'The theory of atomic structure came into being with the discovery of the electron in 1897. . . . [N]o substantial progress was made in the design of atomic models before the commonest part of the atom, the electron, had been chipped off, measured and recognized for what it was.'

6 For more on Nicholson's model, see Maier (1964: 448–461)—who is especially concerned with how this related to spectroscopic findings—and McCormmach (1966).

7 Heilbron (1977) links this tradition to Cambridge, and the Cambridge wranglers in particular. One of its key proponents, as we will see, was Kelvin. Other wranglers included Larmor and Thomson. See also Suárez (2009), who covers Thomson's atomic model and Maxwell's ether. Suárez (2009: 169) uses these examples to motivate his view that 'the hallmark of scientific fiction is *expediency in inference*.'

8 Another development is worthy of a brief mention. Ritz (1908) spotted that the sums of and differences between the frequencies of some spectral lines correspond to the frequencies of other spectral lines.

9 Sometimes Geiger and Marsden are not even *mentioned* when the experiment is (or the experiments are) discussed. Consider, for example, the following online text for a 'quantum mechanics for engineers' course at the University of Colorado at Boulder: 'Rutherford carried out thin gold foil experiments in 1909–1912 in which he passed his α, β and γ particles though films in an effort to determine more about the α, β and γ particles while simultaneously investigating the atoms of thin film itself using these quantum particles as the probes' (http://ecee.colorado.edu/~ecen5016/Chapters/Chapter1.pdf). To the theoreticians go the glory!

10 See also Andrade (1962: 38–39), who writes that even in 1911: 'Rutherford . . . did not seem to regard his theory [of the nuclear atom] as of supreme significance.'

11 Andrade (1962: 38–39) judges that the change in Rutherford's view occurred late in 1912.

12 Dirac (1977: 109) remarks that this is far from unusual:

> [T]he point of view of the historian of science . . . is really a very different point of view from that of the research physicist. The research physicist, if he has made a discovery, is then concerned with standing on the new

vantage point he has gained. . . . He wants rather to forget the way by which he attained this discovery. He proceeded along a tortuous path, followed various false trails, and he does not want to think of these. He is perhaps a bit ashamed, disgusted with himself, that he took so long.

13 Kragh (2012: 5) later adds that Fecher, in 1828, 'suggested a dynamical model of the atom in close analogy to the solar system'.

14 Recall what I said in Chapter 2 about how 'negative' and 'positive' designations are reversible! For more on Weber's work, see Wise (1981) and Assis et al. (2011). Interestingly, and somewhat prophetically, Zöllner wrote in 1876 that: 'The laws developed by Weber about the oscillations of an atomic pair will probably lead to an analytical determination of the number and position of the spectral lines of the chemical elements and their connections with the atomic weights' (Assis et al. 2011: 75).

15 See Heilbron (1977: 60–61) for more detail.

16 For example, as Aaserud and Heilbron (2013: 151) explain:

> The upshot of the council's deliberations was a consensus that, in its straightforward construal, classical physics did not lead to Planck's formula, and, though perhaps not with full agreement, that the needed improvement might be located, and even found, in considerations of atomic structure. . . . The Solvay summit ended on Friday 3 November, the day that Bohr began his weekend visit to Lorrain Smith in Manchester. Rutherford therefore was fresh from the great conclave when he met Bohr. It is hard to imagine that, after learning Bohr's interests, Rutherford did not give him some news of the latest thinking of Einstein, Lorentz, Planck *et al.* about the failure of ordinary physics in and around the atom. The probability is increased by Rutherford's attachment to the Solvay enterprise.

On some key developments that I haven't covered, given my focus on spectroscopy and models of the atom, see Aaserud and Heilbron (2013: 118–125).

17 Heilbron (1977: 69) argues that Nicholson's 'inspired guess' that 'the lines of a series may not emanate from the same atom, but from atoms whose internal angular momenta have, by radiation or otherwise, run down by various discrete amounts from some standard value' was significant in prompting Bohr's interest in explaining spectroscopic phenomena with his atomic model.

18 Earnshaw was yet another wrangler, and a senior one to boot.

19 The situation is eloquently described by Bohm (1951: 38–39):

> According to classical theory, an electron in a given orbit should radiate light having either the frequency of rotation in the orbit, or else some harmonic of their frequency. If, for example, the electron moves in a circular orbit with uniform speed, then only the fundamental should be radiated. In a highly elliptic orbit, however, the particular speeds up a great deal as it approaches the nucleus, and this produces a sharp pulse of radiation, which is repeated periodically. This sharp pulse produces corresponding harmonics in the radiated frequency.
>
> Now, the frequency of rotation depends on the size and shape of the orbit, which are, according to classical mechanics, continuously variable. Hence, there should be a continuous distribution of frequencies in the spectrum, emitted by an excited atom. . . . Furthermore, according to classical physics, if a given frequency v is emitted, then various harmonics of this frequency may also appear, depending . . . on the nature of the orbital motion . . . observed groups of frequencies . . . do not stand to each other in the ratio of harmonics.

20 See Vickers (2013: §3.2.1). I agree with Vickers that the model was not internally inconsistent.

21 Aaserud and Heilbron (2013: 146) add that:

> Here Poincaré identifies a pervasive weakness of the Cambridge school of mathematical physicists. Although they could wield advanced quantitative methods, they had trouble bringing their analyses of the products of their trained imaginations into quantitative confrontation with experiment.

22 Larmor and Thomson were contemporaries. In 1880, the former was senior wrangler and the latter was *proxime accessit* (runner up).

23 Larmor also rated the Zeeman effect as highly significant. In one of the appendices to the book, entitled 'Magnetic Influences on Radiation as a Clue to Molecular Constitution', he wrote:

> The most direct and definite experimental indication towards the intimate structure of a molecule, hitherto obtained, has been the effect of a magnetic field on the character of its free periods of vibration, discovered by Zeeman (Larmor 1900: 341).

24 Bohr made several different derivations, at different points, using his model. Heilbron and Kuhn (1969: 266–283) cover these in considerable depth. The following treatment, however, is loosely based on Bohm's (1951: §2.14).

25 It is remarkable that Fowler (1912) reports experimental results from a *mixture* of helium and hydrogen, but confidently attributes the spectral lines to hydrogen. Bohr's case was helped by a later experiment conducted by Evans (1913), via a request to Rutherford, on helium alone. Fowler (1913) still had doubts initially, but Bohr managed to assuage them; see Fowler (1914: §12–13).

26 For details of Sommerfeld's contributions, see Eckert (2014).

27 This translated from German by Mehra and Rechenberg; for the original, see Gerlach (1969).

28 For more detailed history, see Mehra and Rechenberg (1982: ch. 4).

29 Both Landé and Pauli assuredly followed in the footsteps of Sommerfeld, as both were his doctoral students. The former completed his PhD just before WWI started, whereas the latter completed his in 1921, at the tender age of 21.

30 Kronig (1960: 19–20) recollects:

> Pauli's letter made a great impression on me and naturally my curiosity was aroused as to the meaning of the fact that *each individual* electron of the atom was to be described in terms of quantum numbers familiar from the spectra of the alkali atoms. . . . Evidently s could now no longer be attributed to a core, and it occurred to me immediately that it might be considered as an intrinsic angular momentum of the electron. In the language of the models which before the advent of quantum mechanics were the only basis for discussion one had, this could only be pictured as due to a rotation of the electron about its axis.

31 Kronig (1960: 21) recalls:

> For some reason I had imagined him as being much older and as having a beard. He looked quite different from what I had expected, but I felt immediately the field of force emanating from his personality, an effect fascinating and disquieting at the same time. At Landé's institute a discussion was soon started, and I also had occasion to put forward my ideas. Pauli remarked: "Das ist ja ein gaz witziger Einfall", but did not believe that the suggestion had any connection with reality.

32   In the words of Thomas (1926: 514): 'The value of the precession of the spin axis in an external magnetic field required to account for Zeeman effects seemed to lead to doublet separations twice those which are observed. This discrepancy, however, disappears when the kinematical problem concerned is examined more closely from the point of view of the theory of relativity.'

33   As so often is the case in history, this had been noticed before. Indeed—in line with my earlier findings about how often the same analogies based on everyday experience have been used—the idea that electrons might spin had earlier been proposed in a somewhat different context. Mehra and Rechenberg (1982: 691–692) explain:

> Arthur Compton had assumed in a paper, presented on 27 December 1920 . . . that the electron should be considered as an extended object spinning rapidly around an axis through its centre. . . . Such an electron then exhibited not only an angular momentum, but also a magnetic moment, which accounted for the ferromagnetic properties of metals. However, from the magnitude of the intrinsic angular momentum of the electron—which Compton had assumed to be $h/2\pi$—an important consequence followed. As Compton remarked: 'If an electron with such an angular momentum is to have a peripheral velocity which does not approach that of light, it is necessary that the radius of gyration of the electron shall be greater than 10^{-11} cm'. . . . Kronig did not think of Compton's, or anybody else's, intrinsic rotation of the electron.

See Compton (1921). Mehra and Rechenberg (1982: 692) add in a footnote that 'A year after Compton, Earle Hesse Kennard had proposed, apparently independently, the idea of an electron rotating around its axis'. See Kennard (1922).

34   Mathematically speaking, at least, SR allows for objects that move faster than light. However, most physicists consider this possibility unphysical (or inconsistent with key physical principles). I should add that there is sometimes a misconception that spin 'falls out' of relativistic considerations. This is not the case, as Feynman (1961: 37)—who is accurate on the physics, if not the history—notes:

> The idea of spin was first introduced by Pauli [!], but it was not at first clear why the magnetic moment of the electron had to be taken as $\hbar e/2mc$. This value did seem to follow naturally from the Dirac equation, and it is often stated that only the Dirac equation produces as a consequence the correct value of the electron's magnetic moment. However, this is not true, as further work on the Pauli equation showed that the same value follows just as naturally, i.e., as the value that produces the greatest simplification. . . . [I]t is often stated that spin is a relativistic requirement. This is incorrect, since the Klein-Gordon equation is a valid relativistic equation for particles without spin.

35   Boyd (2002) presents no empirical evidence to this effect. And I suspect many realists would want to distance themselves from the claim. There are significant differences along disciplinary and sub-disciplinary lines, and within disciplines and sub-disciplines over time. These are partly responsible for the wide variation in the way that 'scientific method' is presented in textbooks; see Blachowicz (2009).

36   I don't think there's much wriggle room here. If one wants to say that the Bohr model of the atom counts as 'approximately true', then approximate truth seems too easy to achieve to be philosophically interesting. Quite aside from other considerations, Bohr only had one out of four quantum numbers.

37 The following is based on Bohm (1951: §2.14). Bohr employed many different derivations, at different points. For an extensive discussion of these, see
Heilbron and Kuhn (1969: 266–283).

38 As Muller (2007: 261) notes, 'a majority of the exact equality signs (=) in
most physics papers, articles, and books mean approximate equality (≈).' In
any event, I broadly agree with Vickers (2013: 58) that 'what Bohr was actually committed to in a doxastic sense was . . . Only certain orbits are possible,
and for the hydrogen atom $E_n = \dfrac{hR}{n^2}$ gives *at least a good approximation* to
at least some of them.'

5   Empirical Understanding

In the treatises on physics published in England, there is always one element which greatly astonishes the French student; that element, which
nearly invariably accompanies the exposition of a theory, is the model.
                                                              —Duhem (1954: §IV.5)

As we saw in the last chapter, many Victorian scientists held that models may provide understanding even when they are highly inaccurate or
should not be taken literally. As Heilbron (1977: 42) notes, recall, many
leading nineteenth-century physicists 'considered a theory incomplete
without an accompanying model or analogy, ideally elaborated to the
last detail. Such pictures, they believed, fixed ideas, trained the imagination, and suggested further applications of the theory.'

These scientists therefore endorsed something akin to the notion of
understanding outlined in the first chapter. My case study there focused
on the simple pendulum. I argued that this model helps us to comprehend
the relationship between gravity, length, and period of swing in real pendulums *despite* involving so many idealisations that it is far from being an
accurate representation of any real pendulum. I also argued that a more
accurate model of any given real pendulum would tend to obscure the
insight that this simple model provides us with. That's because increasing
the model's accuracy results in increasing its complexity and narrowing
its scope (in this case and many others).

In this chapter, I will explore in greater depth what kind of an account
of understanding a cognitive instrumentalist might adopt. In order to do
so, I will draw on the work of two key historical instrumentalists, namely
Mach and Poincaré, as well as the contemporary literature on the topic of
understanding.[1] I should like to emphasise, however, that other existing
accounts of understanding are compatible with cognitive instrumentalism; a case in point is the account developed by De Regt and Gijsbers
(2017), which I engage with toward the end of the chapter. In short, the
account of understanding developed in this chapter isn't a *part of* cognitive instrumentalism. It's an articulation of an element thereof.

Broadly speaking, there are two ways to develop an account of understanding that's compatible with cognitive instrumentalism (and thus several other forms of anti-realism). First, one might accept that understanding is reliant upon explanation, but deny that explanation is factive or quasi-factive. Second, one might argue directly that understanding is neither factive nor quasi-factive. This latter route is interesting partly because it's compatible with accepting that explanation *is* factive or quasi-factive, as I did earlier in the book.

The first route is older; a potentially suitable account of explanation was proposed by van Fraassen (1980). The second path has been blazed more recently, by Elgin (2004, 2007, 2008, 2009, 2017), De Regt and Dieks (2005), and De Regt (2009, 2015, 2017).[2] I follow this newer path.

1. Opening Gambits

In the next two sections, I will present my account of understanding. Before doing so, however, I should like to concede some territory to those who take understanding to rest on explanation, or take understanding to be (quasi-) factive in character. I make these concessions to allow that there is something attractive about such views. I will restrict my attention to cases of propositional understanding, as is typical.[3]

First, I grant that understanding *that*, as distinct from understanding *why*, is factive. The two notions are significantly different. For example, many people understand that thunder follows lightning without understanding why thunder follows lightning; indeed, many such people can offer no coherent story whatsoever about how (or why) lightning relates to thunder.[4] As Strevens (2013: 511) puts it, understanding *that* is 'entirely distinct' from understanding *why*. However, this doesn't preclude the possibility that understanding *why* necessarily involves understanding *that*. And I grant that it does, in so far as I accept that in order to understand why p, one must understand that p. For example, I do not think that one can understand why increasing the length of a simple pendulum will increase its period of swing unless one understands *that* this is so. The argument for this is straightforward. Imagine a student who knows (and grasps) all the equations of motion governing the simple pendulum *except* those involving period of swing. (She could be in the process of deriving results for the simple pendulum for the first time.) So she knows (and grasps) that the angular frequency of the pendulum (and hence its frequency of swing) is a function of its length. But she fails to see, because she has not yet considered, how this relates to its period. She is on the cusp of understanding *why* increasing length results in increasing period. But she must notice this first, by deriving and grasping the equation for period, in order to do so: she must understand *that* a particular equation concerning period holds for the simple pendulum.

Some instrumentalists might instead endeavour to deny that *understanding that* is factive. However, the most obvious strategy for so doing, which is to argue that it suffices for a subject to have a psychological relation to a proposition p in order to understand that p, fails. This is because it conflates understanding p (which is not a propositional attitude) with *understanding that p* (which is a propositional attitude). And evidently, understanding 'heat is caloric' doesn't require understanding *that* heat is caloric. A more promising strategy may be to suggest that understanding that requires only the approximate truth of its objects. *Prima facie*, for instance, one is capable of understanding that one is six-foot tall even if one is only approximately six-foot tall. However, allowing this does not appear to make it any easier to resist realism. For example, a realist might nevertheless hold that possessing a scientific understanding of why p requires understanding that q, where q is a scientific explanation of p.[5] In short, the anti-realist loses little by conceding to advocates of factive accounts of understanding, such as Hills (2016: 663), that 'You cannot understand why p if p isn't true'. The anti-realist must instead contest the claim that 'You cannot understand why p . . . if your belief as to why p is mistaken' (Hills 2016: 663).

Second, I grant that if one grasps an explanation of a phenomenon p and grasps all the (non-trivial) consequences of said explanation (as well as that they *are* consequences), then one understands why p. I deny the converse because I deny that in order to understand why p one must grasp an explanation of p (in the factive or quasi-factive sense of 'explanation'). However, this is compatible with holding that understanding why p requires grasping a *potential* explanation of p. 'Potential' can be spelled out in several different ways in this context.[6] But in line with the discussion in Chapter 2, suffice it to say that a potential explanation is like an explanation, save that it may lack truth or approximate truth.

So I accept that if one grasps all the truths (i.e., 'the truth') concerning unobservable entities that bear on some phenomenon (and grasps how and that they bear on it), then one understands that phenomenon. But I think there's a problem with holding that grasping so much is necessary, rather than sufficient, to understand a phenomenon. That's because scientific understanding (why) is then exceptionally hard—if not impossible—to achieve. Such sets of truths are often not mentally accessible, in so far as they are too complex for us to comprehend, involve properties of kinds with which we're not acquainted (because they are not borne by any observable entities), and so forth. Even if our contemporary theories happen to be true, indeed, the demand would be too high. For instance, we would have to accept that we don't understand why any of the planets in the solar system move as they do, in so far as we're not able to take into account the effects of every body with mass (all of which are relevant in the calculations, strictly speaking, in so far as any body with mass exerts a force on any other). We would also have to accept that there's no real prospect of such an understanding.

One potential riposte is that a high degree of understanding is nevertheless possible, even though full understanding is impossible, when it comes to cases such as the movement of nearby planets. However, this is at odds with the way that we normally attribute, and think of, scientific understanding. We take ourselves (or scientists) to understand planetary motions, not just to understand them to a considerable extent (or for all practical purposes). And we don't take that understanding to rely on knowing the positions and masses of all particles emitted by the sun, or even the masses and positions of all the asteroids in the solar system. (We *might* take it to rely on the truth of the assumption that the effects due to such things are negligible. But that is a different matter.)

I should also reiterate that grasping so much might not provide one with the *best possible* understanding of the phenomenon on my view. The reasons for this will soon become apparent.

2. An Anti-Realist View of Understanding I: Mach and Poincaré

In this section, I will argue that Mach and Poincaré implicitly endorsed the view that increasing or achieving understanding is scientifically progressive. I will then draw on Poincaré's thoughts on understanding in order to provide a basis for my own account, which I will develop, with reference to recent work on understanding, in the next section.

As indicated in Chapter 2, Mach thought of scientific progress in an instrumentalist, empiricist, fashion. For example, he declared that: '[i]t is the object of science to replace, or save, experiences, by the reproduction and anticipation of facts in thought' (Mach 1893: 577). And partly as a result, he took scientific discourse about unobservable entities to be valuable only in so far as it helps to achieve this object: '[w]hat we represent to ourselves behind the appearances . . . has for us only the value of a *memoria technica* or formula' (Mach 1911: 49). These passages suggest that understanding why plays no role, even implicitly, in his image of science.

However, Mach (1984: 37) elsewhere wrote that: 'The biological task of science is to provide the fully developed human with as perfect a means of orientating himself as possible.' And this opens the door to scientific understanding having intrinsic or extrinsic value, although Mach did not recognise this. To see why, consider what 'as perfect a means of orientating . . . as possible' would be like.[7] Evidently, it would be empirically adequate. Moreover, it would be simple and thus require minimal use of memory; hence, Mach's emphasis on the significance of economy. But it would also:

(a) be stable—i.e., be resistant to (i) recall failure from memory, and (ii) deletion/loss from memory,
(b) be comfortable and easy to use, and
(c) instil confidence in the user.

Consider the following analogy in support. Imagine you are going on a long hike in the mountains, and have purchased a diligently researched and exquisitely designed map in order to help you to navigate. The map is completely accurate, and has been crafted with simplicity in mind. It depicts no feature that isn't strictly necessary for situating oneself, and is uncluttered and monochromatic as a result. It is also compact, light, and easy to carry.

You set forth on a glorious summer morning, under a clear blue sky. You are delighted with your new map, on the infrequent occasions that it proves necessary to use it. You fail to see how it could be significantly improved upon. So as thick fog rapidly descends later in the day, as you are approaching a series of narrow mountain ridges, you are confident that your progress will not be significantly hindered. Your opinion soon changes, however, when you realise that the map would have been easier to read—that what it represents about the topography would have been easier to grasp—if the contour lines, which are now crucial for navigation, stood out rather more. Perhaps, you reflect, they might have been drawn in a different colour, or even in a range of colours to indicate different heights? This would have made it quicker for you to determine your position each time you looked—and the cumulative boon would have been considerable—although you concede that the map is *sufficient* for you to determine your position.

Matters become rather worse when the gale begins, and you realise how difficult it is to unfold the map and secure it, such that you can read it. The heavy rain, which descends shortly thereafter, is the *coup de grâce*. The map begins to disintegrate due to the elements, and your confidence in making it off the mountains begins to plummet. You're in for a long night.

This analogy illustrates the importance of each of the aforementioned factors in assessing the quality of a tool for orientation. It also strongly suggests that there is a link between understanding and factor (b), namely comfort and ease (and hence speed) of use. But it doesn't relate the other factors to understanding, primarily because of differences between mental and physical representations. To foreshadow the discussion in the next section, however, the stability of (and confidence provided by) a mental representation may depend on the extent to which it provides understanding, whereas the stability of (and confidence provided by) a physical representation is typically independent of the understanding it can facilitate.

Nevertheless, the foregoing analogy might have persuaded Mach to soften his views on theories and models involving unobservable entities, in so far as these may provide sublime *tools for orientation*. Indeed, Mach emphasised the significance of memory and the importance of stability elsewhere in his writing. For instance, he declared that:

> The aim of scientific economy is to provide us with a picture of the world as complete as possible—connected, unitary, calm and not

materially disturbed by new occurrences: in short a world picture of the greatest possible stability.

(Mach 1986: 336)[8]

It is also noteworthy that Preston (2003: 261) construes it to be (at least compatible with) the view of 'the arch instrumentalist, Ernst Mach . . . that the value of natural science lies in . . . impressive predictive/*explanatory* power.'[9] This brings me to Mach's contemporary Poincaré, who suggested that theories and models involving unobservables were valuable, even if they should not be taken literally. More strikingly, he also linked some of the factors discussed previously with understanding. Consider the following two passages:

[Some hypotheses have] only a metaphorical sense. The scientist should no more banish them than a poet banishes metaphor; but he ought to know what they are worth. They may be useful to give *satisfaction to the mind*, and they will do no harm as long as they are only indifferent hypotheses. [emphasis added]

(Poincaré 1905: 182)

[I]ndifferent hypotheses are never dangerous provided their characters are not misunderstood. They may be useful, either as artifices for calculation, or *to assist our understanding* by concrete images, *to fix the ideas*, as we say. They need not therefore be rejected. [emphasis added]

(Poincaré 1905: 170–171)

So Poincaré associated understanding with *mental satisfaction* and *idea fixation*. It is evident that the second item, *idea fixation*, is the same as factor (a). One might call this 'memorability'. It's also plausible that the first item, *mental satisfaction*, is closely related to factor (b). That's to say, it's plausible that a model which mentally satisfies a subject will be more comfortable and easy for the subject to use, *ceteris paribus*, than one that fails to mentally satisfy said subject. I will say more about this in the next section.

For the moment, we can see that Poincaré's discussion suggests the following principle concerning understanding (in which I use 'hypothesis' as an umbrella term to cover theories and models):

If S understands *why* P via a hypothesis H, then:

(i) H provides mental satisfaction (to a significant degree) to S, at least in so far as it concerns P, and
(ii) H is memorable for S.

We may also want to allow for degrees of understanding, given that it's standard to say that some models serve as better vehicles for understanding than others (for psychologists of science as well as philosophers of science).[10] Then Poincaré's comments suggest comparative principles such as:

> If S understands *why* P more with hypothesis H' than with hypothesis H, then:
>
> (i)　H' provides more mental satisfaction to S (concerning P) than H provides, or
> (ii)　H' is more memorable for S than H is.

Such principles provide a useful starting point for developing an anti-realist account of understanding. I turn my attention to this task in the next section.

3. An Anti-Realist View of Understanding II: Modern Insights

There are two matters on which I've intentionally remained relatively silent: (1) what constitutes understanding, and (2) how the factors I've associated with understanding interrelate. In this section, I will tackle these issues with respect to each factor in turn.

Mental Satisfaction (and Grasping)

Let's begin by considering mental satisfaction, which is evidently phenomenological and plausibly also transparent, like joy and sorrow. How might we relate this to understanding? Refining an earlier suggestion by Zagzebski (2001), Grimm (2012: 107) answers as follows:

> There . . . seems to be a kind of first-person immediacy to the experience of "having made sense of the world" that one cannot be mistaken about—even if it turns out that one's way of making sense of things (put differently, the model which one grasps) is fundamentally mistaken. . . . What arguably *is* transparent . . . is what we might call a *subjective understanding* of how the world works—that is, whether one has grasped a model of how the world works that "makes sense".

How does this relate to mental satisfaction? The short answer is as follows. Grimm thinks subjective understanding involves the 'first-person immediacy' of a kind of 'experience'. It is natural to think that the experience involved is (a form of) mental satisfaction. And Grimm (2012: 109) thinks exactly this:

> [T]here is a kind of legitimate *satisfaction* that accompanies our experience of "having made sense of things"—even when this *satisfaction*

is quite low-level, and even when (as it happens) one has only made sense of things from the inside, or relative to one's own experience. [emphasis added]

Grimm contrasts such subjective understanding with 'objective understanding'. Objective understanding involves subjective understanding 'of how the world works' which is also representationally accurate. That's to say, objective understanding of why *p* is factive (or quasi-factive), although it requires (non-factive) subjective understanding of why *p*.

Two comments about this distinction are in order. First, it involves an arbitrary bifurcation, which is liable to lull one into thinking that a subjective account is the only alternative to an objective account. Why not instead say there are several (philosophically interesting) forms of understanding that fall short of being representationally accurate, without being purely subjective? Why not, for instance, posit a form of understanding that involves saving or approximately saving a set of phenomena? Call this *empirical understanding*. It surpasses subjective understanding in so far as it involves a relation to the world, although it does not constitute objective understanding. And an instrumentalist may hold that scientists reliably achieve empirical understanding, while denying that scientists reliably achieve objective understanding (at least when grasping models incorporating unobservable entities).

Second, the fact that *any* potential kind of understanding (why) appears to involve subjective understanding—be it objective, empirical, or what have you—suggests that subjective understanding lies at the core, and is a constitutive component, of understanding (why). It therefore deserves closer investigation, to which I will now turn my attention.

Grimm (2012: 108) claims that subjective understanding has two aspects:

> To begin with, there is the aspect of successfully grasping how the model works—that is, of being able to identify how the various elements described by the model are supposed to depend upon, and relate to one another. . . . In addition . . . there seems to be an aspect according to which the model "makes the best sense" of one's experience, in that it strikes one as the likeliest model, given one's experience. The two aspects are capable of coming apart . . . because one can grasp a model which one does not take to be true.

The first aspect is relatively uncontroversial, and provides the basis for the abilities we associate with understanding, such as the ability to manipulate a model (or representation) that Wilkenfeld (2013) focuses on.[11] The second aspect, however, is more speculative. I hold that it is sufficient to say instead that there is an aspect according to which a model may make *satisfactory* sense of one's experience, or strike one as a *suitable* model given one's experience *relative to the ends for which one intends to use the model.* A key reason is that one may be—and in science, typically is—aware that the best possible model has probably not yet been conceived of (and may even be inconceivable). Consider, for instance, how the models implemented in simulations—used for predictive purposes in climate science, for instance—may be too complex to mentally grasp.

Prima facie, Grimm might respond that he should have written 'likeliest available model', in place of 'likeliest model'. But the difference is non-trivial, in so far as thinking that one has the likeliest available model doesn't entail (or in any sense require) taking said model 'to be true' (as Grimm says), or even to be approximately true. As we saw in Chapter 3, one's probability assignments concerning such items are susceptible to dramatic changes due to new acts of conception. And if one knows this, one might reasonably doubt that there's any significant correlation between subjective probability of such items *at a point in time, in a limited conceptual space*, and the truthlikeness of those items.

Moreover, even granting the psychological reality of an aspect of understanding that involves seeming to be true (or approximately true) it is arbitrary to say that there are no further significant aspects that involve the model seeming to be something else, such as suitable for other desirable ends (which might overlap with those for which it was constructed). One such scientific end is saving the phenomena (or key parts thereof) in a memorable way. Another such end, which those who focus on model-world relations are liable to forget, is improving understanding of a target model; that's to say, sometimes a model can be of use in understanding, and improving one's ability to use, a different model which is suitable for different desirable ends. In each case, aiming for suitable models, or models that satisfice, is typically preferable to searching for models that maximise, in science as elsewhere: there is always an opportunity cost associated with building new models, and the price may be too high when existing tools are good enough—although they manifestly aren't the best possible—for the job at hand. As Kuhn (1962: 76) memorably put it, 'retooling is an extravagance to be reserved for the occasion that demands it'.

In summary, a model can provide mental satisfaction in two distinct respects: in so far as one grasps it; and in so far as one grasps it *and* takes it to be sufficient for a desirable end.[12] For example, I find Bohm's model of the double-slit experiment, discussed in Chapter 2, to be satisfying in so far as I grasp how it relates the trajectory that a particle will take to its initial position. I also find it satisfying because it's a neat way to think about how an interference pattern can be generated by successively firing single particles through the double-slit apparatus. It saves the phenomena—the patterns generated—without appeal to strange events like particles passing through more than one slit at a time, or converting into waves, or to fundamentally chancy laws that do not accord—at least superficially—with everyday experience.[13]

One further issue, namely the role that mental satisfaction plays in inquiry on this account, is worth remarking on at this juncture. Consider Kvanvig's suggestion that (2011: 88):

> [C]uriosity is best understood in terms of a drive or interest or desire for objectual understanding, the kind of understanding which takes an object, rather than a proposition, as its target. . . . It is the perceived achievement of objectual understanding . . . that produces the 'Aha!' and 'Eureka!' experiences that provide closure of investigation into the subject at hand. . . . It is the putting of the pieces of the intellectual puzzle together that does that.

Pay no heed to Kvanvig's emphasis on objectual understanding in this passage, which is a needless distraction; as Kvanvig (2011: 88) notes a little later:

> This approach is sustainable even when inquiry is directed at a single proposition. . . . In limiting though atypical cases, one's inquiry is driven by an initial concern about a single proposition, and what sates the appetite in question for information is a body of information that includes an answer to the initial concern.

Focus, instead, on the issue of closure: focus on the issue of how understanding may affect the flow, or the direction, of inquiry. The account I have offered is compatible what Kvanvig says, in so far as having 'an answer to' a concern, or even possessing an accurate representation, may constitute a desirable end. In addition, the direction a scientist's work takes may be affected by her achieving understanding of a model that she takes to further *other* scientific ends, such as saving some phenomena. Since satisfaction comes in degrees, like joy and sorrow, different degrees of satisfaction may suffice to alter the course of inquiry in different contexts.

Stability/Memorability

This brings us to a second item associated with understanding on the list developed in the previous section, namely stability/memorability. Is it a *part* of understanding why? I answer in the negative because fleeting understanding seems possible, even if it is unusual: there are moments, familiar to those involved in intellectual life, where insight materialises but rapidly evaporates, because of a distraction or a change in mental focus. Despite this, however, memorability is *promoted* by understanding why via a model. Why? The short answer, which I will elaborate on later, is that grasping a model involves spotting connections, and more particularly becoming acquainted with a new *structure*. Being acquainted with a structure gives one a mental anchor for recalling the items that feature therein.

To illustrate the idea, consider again the solar system model of the atom discussed in the last chapter. In order to grasp this model in a scientifically satisfactory fashion, one must grasp it *as* a model; thus, one must grasp to what extent it is intended to be analogous to, or to represent, its target. Hence, one must also independently know something about its target system, namely the atom. One must know that it contains protons and electrons, among other things.[14]

That is to say, one must grasp not only several relations between the sun and the planets in the solar system (or, more precisely, in an idealised abstract model thereof)—'The sun is larger than the planets', 'The planets revolve around the sun', 'The sun is more massive than the planets', and so on—but also grasp what should be taken to be the *corresponding* entities in the atom. For instance, one must grasp that when 'nucleus' is substituted for 'sun' and 'electrons' is substituted for 'planets' in the aforementioned statements, truth will be preserved. (As noted in Chapter 2, these are *structural* resemblances.) One must also grasp that many statements that might be generated by extending the analogy to corresponding items fail to hold. For instance, 'The sun emits light' and 'The nucleus is composed of two kinds of particle' are not items of interest.[15]

In summary, one must have a stock of information about the model, the target, and the putative model-target relation ('mapping information') in order to employ the model, and hence, plausibly, to grasp it *as* a model of something else in particular. But what if one forgets some of this stock? Then one might nevertheless have a basis for recalling forgotten parts—albeit perhaps, imperfectly—by appeal to the aspects one remembers. And here's where information about structure, and structural resemblance ('mapping information') in particular, plays an important role. After learning the solar system model of the atom, for instance, one tends to retain the idea that each system involves things of one kind revolving around, or orbiting, a thing of a different kind. And retaining this improves the memorability of the model (as a model). Naturally, this is compatible with making mistakes in the process of reconstruction, which may or may not be consequential. On the one hand, for example, a secondary school student might mistakenly think that the sun orbits the Earth, and thus compare the Earth to the nucleus. He might therefore take the sun to be a planet, and planets to map on to the electrons. But this needn't cause the student too much trouble. (In effect, he would just take the placeholders in the salient relations to be different on one side, but not on the other.) On the other hand, more consequentially, another student might get worried, when trying to recall the significance of the analogy, about whether the nucleus could be composed of a cluster of different kinds of things, or about whether there are any analogues to moons in the atom.

In summary, grasping structure *promotes* memorability of a model, in so far as it increases the (aleatory) probability of being able to reconstruct it correctly if one forgets aspects thereof. Moreover, it significantly increases the probability of being able to *partially* reconstruct it correctly, even failing that. And partial reconstructions may sometimes be just as good as accurate reconstructions (as the case of the student thinking that the sun revolves around the Earth illustrates). In summary, one can remember a structure without remembering what occupies some, or all, of the nodes occupied in a particular system instantiating that structure.

Comfort/Ease of Use

The penultimate item on our list of factors is comfort/ease of use, which I'll refer to as ease of use, for simplicity's sake, in what follows. Is this constitutive of understanding? Or is it only promoted by, or caused by, understanding?

As a way into finding an answer, recall a key claim that I identified as plausible in the previous section: a model that mentally satisfies a subject will be easier for the subject to use, *ceteris paribus*, than one that fails to mentally satisfy said subject. We might add that two models of the same target may differ in ease of use, for a given subject, in virtue of one providing the subject with greater understanding *why* than the other. That is to say, when understanding *why* is cashed out in terms of grasping (as discussed under the heading of 'mental satisfaction').

As I explained in the previous section, understanding *why* via a model involves more than grasping said model. But for present purposes, we can put this to one side in order to streamline the discussion. That's because two models may afford differential understanding *why* of the same system, for a given subject, because the subject fully grasps one model but only partially grasps the other. Alternatively, the subject might partially grasp each model, but grasp one to a greater extent than the other.

A word on partial grasping is now appropriate. How should this be construed? One simple way in which it occurs is when a subject (fully) grasps some elements of a model, but fails to (fully) grasp others. Consider Bohr's model of the atom. Physics students often take this to involve electrons jumping, instantaneously, between energy levels (and corresponding 'orbitals', or more accurately angular momentum states). But as we saw in the last chapter, Bohr's model does not specify any process by which energy levels change. Bohr (1913) ventures only that the process cannot be described by classical physics.

But how exactly does the degree to which one grasps a model relate to the ease of using said model, *ceteris paribus*? The key to answering lies in Grimm's (2011: 89) recognition that grasping (a scientific model) involves: 'an ability not just to register how things are, but also an ability to anticipate how certain elements of the system would behave, were other elements different'. Let's call these abilities *registration* and *anticipation*. In short, these two abilities are typically compromised—are typically limited—when only partial grasping is present. The more partial the grasping is, moreover, the more the abilities are typically compromised.

As an illustration, consider again the simple pendulum. Upon fully grasping the model, one grasps that the fictional objects of the model swing perpetually. Thus, one *registers* that anything fully represented by the model will swing perpetually. One also grasps, for instance, the equation relating frequency to gravitational field strength, which holds for the fictional objects. Thus, one *anticipates* that the rate of swing of any of the model's legitimate targets will increase when it is exposed to a stronger ('downward') gravitational field, and decrease when it is exposed to a weaker ('downward') gravitational field.

How does this relate to ease of use? In the case of the registered fact, a limitation of the model becomes apparent if one believes that no actual pendulums swing perpetually. Thus, using it becomes easier in the sense that one grasps one of its limitations. (One learns what features of targets not to expect it to capture, with respect also to anticipation.) In the case of the anticipated fact, one can immediately see that a pendulum will swing more slowly on the Moon than it does on Earth (given other elementary true beliefs about the relative masses of the Moon and the Earth, and so on).

Imagine now, by way of contrast, a person who grasps most of the model, but not the two aforementioned aspects. He *might* still be able to use the model to register and anticipate the same things. But this will be a harder task. For example, imagine he *knows* the equation relating frequency to gravitational field strength, but doesn't grasp it. Thus to use the model in order to predict what will happen to a pendulum's rate of swing when it is transferred from Earth to the Moon, he has to plug in some different numbers for g in the equation.[16]

In summary, understanding promotes, but is not constituted by, ease of use.

User Confidence

The final item on the list of factors is 'user confidence', which I'll refer to as 'confidence' in the following. Like the previous two factors, I take it to be promoted by, but not constitutive of, understanding why.

Mental satisfaction promotes confidence in so far as it provides a phenomenological indication that grasping has occurred. It is fallible; but it is perhaps reliable, *prima facie*, in so far as there's a high aleatory probability of grasping being present (to some extent) when mental satisfaction is present (in at least some circumstances). However, the fact that grasping and satisfaction both come in degrees makes matters considerably more complex, and raises troubling questions. For example, how does partial grasping relate to degree of satisfaction? It would be wonderful if degree of grasping were proportional to degree of satisfaction, and degree of satisfaction were proportional, in turn, to degree of confidence inspired. Yet to think so appears wishful. And if the harsh reality is instead that partial grasping can cause a high degree of mental satisfaction, then overconfidence becomes possible. For present purposes, I simply grant that this is so, and note that this is an interesting area for further investigation, which should be empirically informed.[17] My account of understanding does not hinge on confidence in a model due to mental satisfaction always being a good thing, and much of it would survive even if it transpires that mental satisfaction typically causes overconfidence. This would mean that understanding why is not as good a thing as it conceivably might have been, not that it is valueless. That's because overconfidence may perform many of the same functions as (measured) confidence, despite its drawbacks.

Summary

I have proposed an instrumentalist account of understanding—but not explanation—in science. At its core is grasping, which is the basis of subjective understanding. But scientific understanding requires more than subjective understanding, on my view. It requires *empirical understanding*, which also involves saving, or approximately saving, a set of phenomena.

The value of *empirical understanding*—as distinct from merely having the ability to save (or approximately save) the phenomena—lies in its effects. It makes it easier for a subject to remember and use a model. It also tends to improve the user's confidence in the model (as a means towards desired ends). So the presence of empirical understanding contributes to scientific progress—or aids in achieving 'the task of science'—on traditional instrumentalist views about science, such as Mach's.

4. Comparison With the Account of Understanding as Effectiveness

An alternative anti-realist view of understanding—an understanding of understanding that is compatible with cognitive instrumentalism—has recently been proposed by De Regt and Gijsbers (2017). I will now present this and argue that the account developed in this chapter is preferable in several respects.

De Regt and Gijsbers's central claim is that understanding can only be gained from a scientific device, such as a model or theory, which is *effective*. 'Effective' is a technical term, which they define as follows:

> A device is effective just in case the device is usable by the scientist and using it reliably leads to scientific success. So an effective device must be both intelligible and successful; in a quasi-formula:
>
> effectiveness = intelligibility + reliable success.
>
> (De Regt and Gijsbers 2017: 55)

Let's deal with the two parts of *effectiveness* in turn. Intelligibility is defined in terms of usability, or ease of use: 'a theory T [or other representational device] is intelligible for a scientist S if . . . it has qualities that facilitate its use by S' (ibid.). Success involves performing 'the core tasks of science', which are threefold: 'making correct predictions', 'guiding practical applications', and 'developing better science' (ibid.). Thus, a 'successful representational device is one that performs well—that is, reliably leads to success—on one or more of these scales of appraisal' (ibid.). The forms of success are related, such that, for example, the final task cannot be furthered without furthering at least one of the other two at the same time.

This brings me to my criticisms of this view. First, as I argued in the previous section, ease of use is not constitutive of understanding. It follows that 'intelligibility' in De Regt's sense is not. Second, it is possible for a scientist to be able to use a device—such a model—in such a way that leads to reliable success *without* anything deserving the name 'understanding'—and notably without 'subjective understanding' (Grimm 2012) or grasping—being present. Consider developing a complex model for a computer simulation, like a climate model. The scientist uses the model in the simulation. And the use might reliably lead to accurate predictions (in a significant class of scenarios). Similar examples can be constructed involving concrete models, like the Phillips model of the economy (and others discussed in Chapters 2 and 3). Do we want to say that merely by developing this model and being able to use it to reliably make predictions about the economy, Phillips came to understand a lot about the economy? Do we want to say, even more incisively, that if we find an alien device we can easily use to make excellent predictions concerning a class of phenomena—such as the next day's weather—then we therefore have new understanding about those phenomena? That is, even if we have no idea, whatsoever, about the workings of the device? De Regt and Gijsbers (2017) might reply that they only propose 'a necessary condition for achieving understanding'. I acknowledge this. However, I take the prior examples to suggest that grasping is a missing (necessary) element in their account.

A related problem with the account of De Regt and Gijsbers (2017) is that it involves built-in assumptions about what success consists in—about what 'the core tasks of science' are. But a realist will dispute the account on those grounds. My approach in this chapter has been different. I have argued that something which anti-realists and realists can agree on the existence of—namely, subjective understanding—contributes to the achievement of scientific ends that cognitive instrumentalists take to be important.

5. Conclusion

There is a form of scientific understanding, *empirical understanding*, which goes beyond subjective understanding but falls short of objective understanding (or any approximation thereto). Moreover, achieving such empirical understanding—above and beyond improving our ability to save the phenomena—may be held to be scientific progressive (or valuable) from an anti-realist (or instrumentalist) perspective. The roots of this idea lie in the instrumentalist tradition, and can be found in the work of Poincaré and (to a lesser extent) in the work of Mach.

Notes

1 Understanding is now a much hotter topic than it was when I began writing this book in 2012. Over the past two years alone, the following have appeared: Dellsén (2016, Forthcoming); De Regt (2017); Elgin (2017); Grimm et al. (2017); Hills (2016); Khalifa (2017); Lawler (2016); Rancourt (Forthcoming); Reutlinger et al. (Forthcoming); Rice (2016); and Stuart (2016). This is not an exhaustive list.

2 For example, De Regt and Dieks (2005: 167) note—in a paper based on a series of talks that began in 1998—that their: 'conception of scientific understanding is sufficiently broad to allow for the possibility that a theory is understood without a realistic interpretation. . . . Therefore, the fact that scientific understanding is an aim of science does not entail scientific realism.' Elgin (2004: 122) argues, more boldly, that: 'Idealizations may be far from the truth, without thereby being epistemically inadequate. . . . There is no reason to think that in general the closer it is to the truth, the more felicitous a falsehood.'

3 A challenge to the view that all understanding is propositional in character stems from Lipton (2009), who claims that one may understand why without being able to articulate said understanding why. Strevens (2013: 512) responds by arguing that one may be able to grasp propositions without being able to express them (and that said propositions might nevertheless be expressible by someone). It is not necessary to take sides in this debate, however, to acknowledge that a significant subset of scientific understanding is propositional in character.

4 Following Khalifa (2012), I allow that understanding why may be spoken of as 'understanding how' in some contexts. But I take there to be no fundamental difference between the two notions.

5 This would involve a weakening of the view championed by Strevens (2008: 3), who takes 'scientific understanding to be that state produced, and only produced, by grasping a true explanation.' (And note that: 'To grasp that a state of affairs obtains is to understand that it obtains' [Strevens 2013: 511].) In effect, it is to substitute 'true' with 'true or approximately true' in this statement. An ardent realist might resist such a move by invoking degrees of understanding, and associated degrees of understanding that; she might concede that a high degree of understanding is possible in the presence of an approximately true explanation, while maintaining that this would fall short of complete understanding.

6 Strevens (2013) suggests that a potential explanation is an explanation that satisfies the *internal* conditions for a correct explanation. On Hempel's deductive-nomological account of explanation, for instance, any potential explanation of p would entail p, and employ a (putative) scientific law in order to do so. Consider, by way of contrast, the view of Bokulich (2009: 104) on which 'fictions [can] . . . carry explanatory force, and correctly capture in their fictional representation real features of the phenomena under investigation'. In short, Bokulich holds that there are explanatory fictions.

I grant that fictions may only be *potentially* explanatory at best. But I hold that they may serve as vehicles for empirical understanding, as I subsequently explain.

7 Mach plausibly had possibility in practice in mind, although I discuss possibility in principle in order to make the discussion more concise. Suffice it to say that when I write of the perfect means as empirically adequate, I might instead have written 'as empirically adequate as possible', and so forth. A complication arises because such virtues are not independent—for example, maximising simplicity may involve sacrificing empirical adequacy—but this doesn't bear on the point I make in this passage.

8 Consider also the following passage:

> What guarantees to primitive man a measure of advantage over his animal fellows is doubtless only the strength of his individual memory, which is gradually reinforced by the communicated memory of forebears and tribe. (Mach 1976: 171)

I have argued, *inter alia*, that Mach was mistaken to link stability *solely* with economy.

9 Preston (2003) doesn't say precisely why he mentions explanatory power, although he rightly criticises Popper's (1962) presentation of instrumentalism as involving a search for utility, unless utility is understood to embrace 'epistemic (albeit not truth-directed) factors' (Preston 2003: 265). Nonetheless, Preston presumably had in mind what I earlier called *potential explanations*, and may very well also have considered the centrality of *orientation* in Mach's views about the value of science.

10 For example, Gentner and Gentner (1983) examined how novices in physics reasoned when using two different mental models for electricity flow, namely moving crowds and water flowing in a pipe. They found the former to be superior for thinking about the effects of resistors, and especially resistors in parallel. In short, as Nersessian (2002: 146) emphasises: 'analogies . . . are generative in the reasoning processes in which they are employed.'

11 I agree with Wilkenfeld (2013: 997) that 'understanding x is somehow related to being able to manipulate x' and that 'understanding is a mental phenomenon', but do not agree that understanding should be *reduced* to an ability to manipulate a *mental representation* in the way he specifies. That's to say, I don't accept that a 'thinker T understands object o . . . if and only if T possesses a mental representation R of o that T could . . . modify in small ways to produce R ', where R ' is a representation of o and possession of R ' enables efficacious . . . inferences pertaining to, or manipulations, of o' (Wilkenfeld 2013: 1003–1004).

I don't accept this because understanding why sometimes involves seeing that an empirical relationship follows from a model, via a mathematical derivation, without being able to *mentally* modify said model in order to see that said relationship holds. Consider again the simple pendulum, as a case in point. When one has worked through the proper derivation, which doesn't involve any modification, one can see that increases in length result in increases in period (*ceteris paribus*). One then gains a (legitimate) sense of understanding why period increases as length increases (and an associated ability to reason counterfactually about how changes in the latter will affect the former). Nothing depends on being able to modify a particular representation so that length increases, and then (mentally) seeing that period increases.

This problem might arise partly because Wilkenfeld focuses on understanding why concerning objects, rather than understanding why in general, which can have other targets, such as features of objects.

12 A minor refinement might be appropriate: one might merely take the model to be *necessary* for achieving the desirable end, or necessary or sufficient for moving closer to the desirable end.

13 In so far as one might find the end of truth (or approximate truth) desirable, moreover, this account of mental satisfaction *subsumes* Grimm's.

14 Frigg and Hartmann (2012: §3.2) puts this as follows:

> Once we have knowledge about the model, this knowledge has to be 'translated' into knowledge about the target system. . . . Models can instruct us about the nature of reality only if we assume that (at least some of) the model's aspects have counterparts in the world.

15 The distinction here is between negative analogies and positive analogies, as discussed in Chapter 2. Something akin to neutral analogies will also be present.

16 As Sober (1975: 67) emphasises, recall: 'Mathematical equations do not merely enable us to calculate the value of one variable given the values of the others. They also characterize the relationships between variables and enable us to compute how any variable changes relative to the others.' Thus one can know and be able to use an equation to calculate the value of a given variable in any particular case, without spotting how it characterises the relationship between the variables it involves.

17 See, for example, Ylikoski (2009), who engages with psychological literature on the illusion of explanatory depth which is also treated in the final chapter, and concludes that 'The simple reliance on the sense of understanding is not a sufficient criterion.' It does not follow that this sense is not a necessary criterion. Nor does it follow that it isn't indicative of understanding in some significant contexts. It should be added that Ylikoski holds that 'understanding is an ability to give explanations'. If one doubts this, as I do, then the relevance of the psychological research on the illusion of explanatory depth is questionable.

6   Objections and Comparisons

In the previous chapters, I presented and argued for cognitive instrumentalism. I also used the history of science to strengthen the case for this view, and developed an account of empirical understanding that could feature therein. In the present chapter, I take on two remaining tasks.

First, I defend cognitive instrumentalism against a line of attack that threatens most, if not all, forms of instrumentalism. This involves questioning the tenability and significance of the distinction between the observable and the unobservable. Second, I explain how cognitive instrumentalism differs from three key alternatives in the current literature, namely constructive empiricism, structural realism, and semirealism. I also present reasons for preferring cognitive instrumentalism to those positions.

1. The Observable-Unobservable Distinction

The distinction between the observable and the unobservable is central to many alternatives to scientific realism: traditional instrumentalism (as championed by the logical positivists, and reflected in Ayer [1936]), constructive empiricism (van Fraassen 1980), structural realism (Worrall 1989b), and perhaps even semirealism (Chakravartty 1998). It also plays a central role in the theses advocated in this book. In the first chapter, for instance, progress is said to involve saving and understanding *the phenomena*. And property instrumentalism, which is presented in the second, involves the idea that we cannot fully comprehend (and discuss) an unobservable entity if it possesses unobservable properties.

But now imagine that everything is observable in principle. Then the truth of the latter thesis as a claim about the unobservable *in practice* is no significant threat to realism, provided that more and more things are likely to become observable in practice over time. Similarly, 'saving and understanding the phenomena' in the former thesis would amount to 'saving and understanding all physical things' *in principle*, and might eventually amount to doing the same in practice.

It is thus of little surprise that realists have paid considerable attention to the distinction between the observable in practice and the unobservable in practice, and have argued that the set of the former tends to expand considerably, whereas the set of the latter tends to contract considerably as a result, as science progresses.

There are two main ways, both explored by Maxwell (1962), in which something unobservable (in practice) might become observable (in practice). ('Observable' and 'unobservable' refer to the 'in practice' notion in what follows, unless otherwise specified.) On the one hand, we might develop new instruments. On the other, on the assumption that (at least some) observation is theory-laden, we might develop new theories. In what follows, I'll consider each way in turn (although it will emerge that they are deeply connected), and argue that each has significant limits.

Beforehand, however, I should like to address a couple of potentially distracting, but relatively insignificant, issues. First, as noted in Chapter 2, 'observable' and 'unobservable' are vague terms. But so are everyday terms like 'orange', 'red', 'smooth', 'river', 'snow', and 'round'. And realists do not typically object to the use of such terms. Among other reasons, this is because a language devoid of vague terms but with sufficient scope and simplicity to be usable in science, even if not in our daily lives, is difficult, if not impossible, to conceive of. Moreover, it may even be open to us, following Williamson (1994), to maintain that there is a fact of the matter about whether a given entity is observable even when we are not able to determine this.

Second, some authors on observability, such as Churchland (1985) and Musgrave (1985), consider cases where new persons, or kinds of person, are admitted into the scientific community. These might be aliens with x-ray vision, for instance. Wouldn't the inclusion of these in our community of inquirers extend the range of the observable? Testimony-related concerns are plausibly no different in this scenario than they are in current human science. If such aliens are admitted into the scientific community, then their testimony will be treated just like a human scientist's. So what extra would the aliens bring? The answer appears to be simple; they'd bring new instruments, albeit 'natural' instruments with which they (*qua* alien persons) were biologically connected. Thus, I propose to treat our own sense organs—our eyes, our ears, and so on—as instruments. This raises some interesting questions that we won't be able to tackle, such as how such instruments should be individuated. But we can work with the rough view that they take inputs and give outputs, and that those outputs eventually go into something that's not an instrument (for observation), such as the brain (or parts thereof). This 'rough and ready' characterisation is reasonable, in the present context, because allowing that parts of the human body are instruments is a significant concession to realists. To appreciate this, consider how an artificial eye might thereby be considered to be on a par with—or even to surpass—a biological eye. The notion that natural instruments are superior to non-natural instruments, either typically or in general, has effectively been abandoned.

We are now in a position to consider the two key ways in which the line between the unobservable and the observable may shift.

1.1 Instruments

Maxwell's (1962: 3) target was instrumentalism of a traditional semantic variety:

> That anyone today should seriously contend that the entities referred to by scientific theories are only convenient fictions, or that talk about such entities is translatable without remainder into talk about sense contents or everyday physical objects, or that such talk should be regarded as belonging to a mere calculating device . . . such contentions strike me as so incongruous with the scientific and rational attitude and practice that I feel this paper *should* turn out to be a demolition of straw men.

However, cognitive instrumentalism does not involve any semantic theses of such generality. It is more modest, in so far as it takes discourse to be literally construable provided that it appropriately connects with experience, in the properties and analogies it involves, as explained in Chapter 2. Thus, it is not one of Maxwell's straw men.

Nevertheless, Maxwell's (1962: 7) argument that there is a continuum between the observable and the unobservable is pertinent, in so far as it calls into question the broader significance of the observable-unobservable distinction:

> It really *does* 'seem awkward' to say that when people who wear glasses describe what they see they are talking about shadows, while those who employ unaided vision talk about physical things—or that when we look through a window-pane, we can only *infer* that it is raining, while if we raise the window, we may 'observe directly' that it is. The point I am making is that there is, in principle, a continuous series beginning with looking through a vacuum and containing these as members: looking through a window-pane, looking through glasses, looking through binoculars, looking through a low-power microscope, looking through a high-power microscope, etc., in the order given. The important consequence is that, so far, we are left without criteria which would enable us to draw a non-arbitrary line between 'observation' and 'theory'.

Maxwell (1962: 8–9) aims to show that the distinction between the observable and the unobservable has no *ontological* significance:

> [I]s what is seen through spectacles a 'little bit less real' or does it 'exist to a slightly less extent' than what is observed by unaided vision? . . . Although there certainly *is* a continuous transition from observability to unobservability, any talk of such a continuity from full-blown existence to nonexistence is, clearly, nonsense.

Cognitive instrumentalism does not suggest otherwise. And I grant that unobservable entities exist, and that we can understand *some* of these in their entirety. But I also believe there are other unobservable entities that do not possess only observable properties, and that we cannot fully comprehend (even by the use of analogies). I also suspect that there are unobservable entities that we cannot comprehend at all. (Belief in these propositions is not part of cognitive instrumentalism; it is compatible with cognitive instrumentalism.) To be confident that this is wrong—that everything is comprehensible—is bold indeed.

I also join van Fraassen (1980, 1985) in taking the distinction between the observable and the unobservable to have *some* epistemological significance.[1] And I would add that it has considerable *practical* significance. I will now explain how so, before going on to consider potential realist objections.

My fundamental premise is simple. We are, for the time being at least, stuck with our natural instruments. And by that, I mean that we have no choice about whether to use them or not, in investigating the natural world. When I look through a microscope or telescope, I use my eye no less than I do when I look at the desk at which I sit. When I listen to the clicks of a Geiger-Müller counter, I use my ears no less than I do when arguing with my wife about who should use the study in which the aforementioned desk lives. And so on. Indeed, many artefactual instruments are *designed* to couple with our natural instruments; binoculars are an excellent case in point.

Moreover, I take my hearing to be defective in so far as I have unilateral tinnitus. I take it to be defective with reference to how I take a properly functioning auditory system to behave, on the basis of my past experience and testimonial and physical comparisons with other humans (and relevantly similar animals). (Assume the tinnitus under consideration is indisputably caused by changes in the ear *qua* instrument, e.g. by cilia damage, although it's possible for tinnitus to have other, e.g. neurological, causes. This avoids concerns we touched upon earlier, about how to individuate instruments.) In general, indeed, the effectiveness of corrective visual or auditory aids is measured with reference to how we take satisfactory or optimal natural instruments to function. Thus, an optometrist might ask you to read aloud letters of different sizes, at a specific distance, in a well-lit environment. Similarly, an audiologist might ask you to click a button whenever you hear pure tones of differing frequencies, while varying the intensity thereof. There are no better alternatives.

One might try to run an evolutionary argument, at this juncture, to the effect that our natural instruments have been 'selected for' in such a way that they accurately represent reality (when they're properly functioning).[2] First, however, this is to be misled by seductive metaphors about 'increasing fitness' or 'becoming well-adapted'. For adaptation towards the optimal is neither guaranteed, nor made highly probable, by the evolutionary process.

Allow me to illustrate why. Consider a population, and assume, for simplicity's sake, that each organism therein produces two offspring with identical attributes (to its parent) if it reaches a particular age. Thereafter, it can reproduce no further. Imagine also that the reproduction is asexual, and that no mutation occurs. But allow that all genetically possible organisms of the kind under discussion exist in the initial population. (This is just a thought experimental trick to simplify matters; that is, to avoid the need to consider mutation over time.)

Some organisms die before they are able to reproduce, because they possess particular attributes (or clusters of attributes), or lack particular attributes (or clusters of attributes). Others survive. But those that survive are not necessarily resistant to being 'selected against' by the environment, even assuming that this remains the same except in so far as the population changes. For one thing, changes in the population can directly impact survival prospects (and hence selection can occur at the group, and not just the individual, level).[3] For another, clusters of attributes might be necessarily connected (e.g. because of how genetic code is stored) in such a way that some combinations thereof, such as optimal combinations from the point of view of the individual, are not possible in offspring.

Consider now a more specific scenario, involving rabbits of two different types, in order to see the point. Those of the first type are fast and nimble, but have white stripes under their tails. Those of the second type lack the stripes, yet are slower and less agile. The types are otherwise highly similar. They face only one predator, the stoat, in their environment. The stoat is rather fast, relies on vision to hunt, and is not terribly smart. So when the rabbits flee due to their presence, the stoats pursue those with the white patches underneath their tails, which they always find easy to visually track, although these escape, in virtue of their agility and speed, about half of the time. Over time, the rabbits with these attributes—those of the first type—diminish in number. They are eventually wiped out by the stoats. This is terrible news for the remaining rabbits. They turn out to be easier prey because they are almost always caught, when they are successfully visually tracked while fleeing, and they are successfully visually tracked about three quarters of the time. (If the stoats had been smarter, they would have learned earlier that hunting these rabbits was a more successful strategy.)

Introducing mutation (along with sexual reproduction, if desired) does not alter matters. For one thing, mutation is typically limited, in so far as some combinations of attributes are inaccessible (or just highly improbable) by mutation. But even if the possibilities due to mutation were unlimited, it is implausible that adaptation would become so. This is emphasised by Barton and Partridge (2000: 1083), who conclude:

> Although much of the emphasis . . . has been on limits to selection set by the nature of the phenotypic variation that can be generated ('developmental constraints'), population genetic constraints may be just as, or even more, important. There is clear evidence that the rate and extent of evolutionary change can be constrained by (1) lack of fit intermediaries leading to an optimal phenotype, (2) limits inherent in the process of selection, set by the number of selective deaths and by interference between linked variants, (3) limits due to evolutionary forces such as drift, mutation and gene flow, and (4) limits caused by conflicting natural selection.

For the sake of argument, let's also, second, imagine that a naïve view of evolution, on which progress to optimality is inevitable, is correct. Would it be optimal to be very good at knowing how the world is at the fundamental level? And, in particular, why would it be any better than being able to save the phenomena in an economical fashion? Imagine two tribes, the Knowers and the Savers. The Knowers are experts at discovering the fundamental structure of the universe, whereas the Savers are experts at saving the phenomena. What survival advantage could the former have over the latter? The Savers know how to intervene efficiently in order to achieve the observable effects that they desire—how to avoid death, to treat symptoms of disease, and so forth. So how could the Knowers do any better? It seems plausible that they might do considerably worse, in so far as knowing the fundamental structure of the universe doesn't guarantee— to revisit a topic discussed at length in Chapter 1—being able to solve practical problems quickly and effectively.

So there's a significant sense in which we're stuck with our natural instruments, at present, although there's little reason to believe that these instruments are optimal when it comes to representing reality (or that they accurately represent reality, to the extent that they do represent it). I emphasise 'the present' because I accept that what's observable is community, and technology, relative. (This is not a concession. Any position that denies this relativity is inadequate. What science can achieve, or can reasonably be expected to achieve, depends on the cognitive and material resources of its practitioners.) But I do not want to speculate about how exactly our community, or our technology, will change in the future. It may transpire that the position I offer collapses into realism (of one variety or another) because of the way that instruments develop. But that's a contingent, empirical, matter.

Nevertheless, it is difficult to see how we would come to renounce interest in our natural instruments altogether. Imagine that we had the ability to replace these. It is plausible that we would only take the replacements to be clear improvements if they performed the same functions as our natural instruments, *inter alia*. But imagine that they didn't perform the same functions. How would we determine if they were *better*? Presumably, as a result of trying them out, navigating the world with them, and seeing how successful we were in achieving our ends (and perhaps even how enjoyable the experiences were); in effect, we would venture into a new empirical world, and try it out. But would increased success in achieving our ends, while in that world, entail greater *veridicality* of perception (howsoever that is to be construed)? The answer lies in the negative. Increased veridicality could even be a curse rather than a boon. Too much information could be distracting, or require too much processing time. (Every processor has its limits.) And information might be available in a form that made it harder to use than it was previously. Imagine, for example, that the colours of visible objects in an implanted camera 'field of vision' were represented numerically. Discerning differences in shades would be easier, but grouping items of the same colour, according to current conventions, might take considerably longer. A riposte might be that such conventions would cease to be of value or use. However, this is unclear. Consider trying to grasp why people thought old paintings were beautiful, for instance.

In essence, then, my position on basic instruments has a pragmatic underpinning. I don't see much point in questioning what we're stuck with, and I think there's a deep sense in which we don't have a choice if we're set on inquiring (and doing science).[4] Realists will tend to agree with this, even if they're more inclined to think that what we're stuck with is good when it comes to finding out truths. I think such instruments are the best *we have*, by way of contrast. That doesn't make them good in an absolute sense. I fail to see how we'd ever be in position to determine whether they are.

Let's now return to the idea that many instruments used in science—instruments such as lens and mirror telescopes—are used *in series* with our natural instruments; or, as Hooke (1665: Preface) puts it, '[operate by] the adding of artificial *Organs* to the *natural*'. Putting 'equipment' in series introduces new opportunities for error (even if it also potentially introduces new opportunities to observe or detect). An eye and a telescope may each be working perfectly, but may not interface satisfactorily; for example, the latter may not be suitably focused. An oscilloscope may be functioning properly without being properly calibrated. And so forth. (Similar problems may arise in attempts to use instruments in parallel. One might have strabismus or amblyopia, for instance.) This gives a basis for some caution. That is, although it should be admitted that instruments in series may be more reliable than any of their components in isolation in some contexts (as we can establish by appropriate comparisons, as in cases of glasses or hearing aids). So why not think that scientific instruments are typically, or can in many circumstances be determined to be, reliable in this way? I will address two distinct arguments to this effect.

The first, which is the weaker of the two, is the 'argument from continuity' offered by Lord Quinton. The idea behind this is that we can initially make useful comparisons between what we see with our unaided senses and with scientific instruments, and then compare what's seen with the instruments in the aforementioned circumstances with what's seen with the instruments in situations where direct recourse to unaided senses isn't available. Or as Quinton (1973: 301–302) puts it at greater length:

> An issue of principle arises only with the objects apparently revealed to our perception when our senses are assisted by various sorts of instrument. The detailed structure of a snow crystal that we see under a magnifying glass is something we should ordinarily regard as having been observed. Is this a legitimate step? What counts in its favour is the fact that all the features of things that are observable without this modest kind of instrumental assistance are still observed with it, along with some other features as well. But once we admit that a thing can be literally observed with a magnifying glass there seems no point at which we can reasonably say that we are observing, not the thing itself, but its effects as we move along the series of ever more refined and sophisticated observational aids: from magnifying glasses to microscopes and from ordinary microscopes to electron microscopes with vast powers of magnification. The argument from continuity applies even to the latter. The properties and constituents of the specimen that are visible without assistance are all seen through the electron microscope at the lower levels of magnification, although greatly enlarged. As the magnification increases some of the detail that was observed at the preceding stage is still there to be seen.

The problem with this argument is that scientific instruments typically generate artefacts, such that their outputs require interpretation. A look to the history of science supports this view. For example, Hooke (1665: Preface) notes, concerning things seen through compound microscopes:

> [O]*f these kind of Objects there is much more difficulty to discover the true shape, then of those visible to the naked eye, the same*

> *Object seeming quite differing, in one position to the Light, from*
> *what it really is, and may be discover'd in another. . . . [I]t is exceed-*
> *ing difficult in some Objects to distinguish between a* prominency
> *and a* depression, *between a* shadow *and a* black stain, *or a* reflec-
> tion *and* whiteness in the colour. *Besides, the transparency of most*
> *Objects renders them yet much more difficult than if they were*
> opacous.[5]

Another excellent way to counter the argument for continuity is to look
to what scientists with expertise in contemporary microscopy say about
what this involves:

> Microscopy is the study and interpretation of images produced by a
> microscope. "Interpretation" is the key word. . . . It is important to
> remember that microscopy is not simply a matter of magnification
> and making objects larger, but of resolving features that may not have
> been seen before. Much microscopical analysis is subjective. Images
> cannot be interpreted intuitively on the basis of nonmicroscopical
> experience but require highly skilled microscopists with knowledge
> and practical experience of the materials they are examining and an
> 'eye' for their subject. Correct interpretation can be achieved only
> when one has a thorough understanding of all the factors that influ-
> ence the final image. These factors include instrumental effects, speci-
> men preparation techniques, and microscope specimen interactions.
> Three-dimensional objects projected onto a two-dimensional image,
> such as is the case with scanning electronmicrographs, can be par-
> ticularly confusing and difficult to interpret. Images of surface fea-
> tures can be illusory.
>
> (Moss and Groom 2001: 150)

To stack the deck in favour of the realist, let's continue to consider microscopes. Let's imagine that what goes for these will 'carry across' to cases
where it seems clearer, *prima facie*, that only something like 'detection'
could be occurring: cases such as neutrino detection (at SNO), or gravitational wave detection (at LIGO).[6] I only make this concession temporar-
ily, for argument's sake. For example, I do not concede that LIGO makes
the observation of gravity waves possible, let alone that LIGO enabled
the observation of a collision between two black holes (as Abbott et al.
[2016] claim). Several *explicit inferences* are required in order for such
'observations' to be made, or more accurately for such existential claims
to be derived. Differences in the time it takes to receive laser signals are
inferred to be due to deflections of mirrors, in each LIGO installation.
Some such (inferred potential) deflections are also 'filtered out' as being
irrelevant. (Noise comes from several sources, in different frequency
ranges: seismic, thermal, and shot.) Only when signal change patterns in

each installation are similar *after processing*, over the same period, may the presence of a gravitational wave be inferred (in the absence of other possible explanations appearing probable). It's a further step to infer the event responsible for the similar (processed) patterns. Such long chains of inference aren't present when everyday observations are concerned. And the length of the chain is significant in so far as errors can creep in at each inferential step.

We can now move on to consider the second argument, from Hacking (1981), which might be called the argument from consilience. Hacking prepares this by arguing 'you learn to see through a microscope by doing, not just by looking' (1981: 136), and denying that theory is important in the use of a microscope:

> It may seem that any statement about what is seen with a microscope is theory-loaded: loaded with the theory of optics or other radia-tion. I disagree. One needs theory to make a microscope. You do not need theory to use one. Theory may help to understand why objects perceived with an interference-contrast microscope have asymmetric fringes around them, but you can learn to disregard that effect quite empirically. Hardly any biologists know enough optics to satisfy a physicist.
>
> (Hacking 1981: 137)

My response to this is as follows. First, microscopes generate images. As van Fraassen (2001: 157) puts it:

> The success of that family of instruments (microscope, electron microscope, radio telescope) derives in part from the possibility of representing their products as images of real things existing indepen-dently of any relations to those instruments. *But their products are images*; they are optically produced, publicly inspectable images. It is these images that are like the rainbow (they cannot themselves be represented as independent things).[7]

Second, these images require interpretation. (Thus, to put matters in terms of the prior discussion of LIGO, there are inferential steps required to map the image, or parts thereof, on to something else.) So the fact that one doesn't need optics to use a microscope—which Hacking notes— doesn't show that one doesn't need *any* theory to use one effectively, or to interpret the images produced. The experts in microscopy, quoted previously, are clear on this: 'Correct interpretation can be achieved only when one has a thorough understanding of all the factors that influence the final image.' So *theories concerning microscopic images* are required, even if theories about how to construct microscopes, or how microscopes work, are not.

We now come to Hacking's argument proper. The basic idea behind this is that similar images, or parts of images, may sometimes be obtained from different processes:

> Two physical processes . . . are used to detect the bodies. These processes have virtually nothing in common between them. They are essentially unrelated chunks of physics. It would be a preposterous coincidence if, time and again, two completely different physical processes produced identical visual configurations which were, however, artefacts of the physical processes rather than real structures. . . .
>
> Note that no one actually produces this 'argument from coincidence' in real life. One simply looks at the two (or preferably more) sets of micrographs from different physical systems.
>
> (Hacking 1981: 144–145)

Hacking overstates the case by writing of 'identical visual configurations', for reasons we've already covered; interpretation is the order of the day, and one would only expect to see *similar* visual configurations, or configurations that *might be interpreted as representing the same* 'real structures'. Moreover, we should ask what would lead Hacking to think it would not be 'a preposterous coincidence' if 'two completely different physical processes' were to provide similar representations of 'real structures'? The only reasonable answer appears to be to appeal to the details of the scientific theories behind the construction of the instruments; that's to say, to make it clear that the objects of discussion are 'two completely different physical processes' *that are sufficient for achieving similar representational ends according to particular scientific theories and models*. But we saw earlier that Hacking downplays the significance of theories (and thus models) in thinking observations are faithful. So a tension develops in his position, and this sets the stage for the discussion of theory-ladenness in the next section.

In summary, van Fraassen (2001: 155) is correct that: 'The instruments used in science can be understood as not revealing what exists behind the observable phenomena, but as creating new observable phenomena to be saved.' However, one should be wary of generalising about instruments. I don't think one should deny that scientific instruments can *sometimes* reveal 'what exists behind the observable phenomena'—or, as I'd prefer to put it, that scientific instruments sometimes make it possible to observe things that were not previously observable. Moreover, where exactly one draws the line between those that can and those that cannot 'reveal more' is of little significance in assessing the position I outline. (It's important only that there's a significant class of instruments at present, like the aforementioned SNO and LIGO, that don't allow observation.) So you and I might disagee on which instruments render things observable—and even on how exactly the line between the observable and unobservable is liable to shift—while agreeing that cognitive instrumentalism is correct (or at least approximately so). Van Fraassen (2001: 162–163) similarly concedes:

> I really don't mind very much if you reject this option [from the last quotation] for the optical microscope. I will be happy if you agree to it for the electron microscope. For optical microscopes don't reveal all that much of the cosmos, no matter how veridical or accurate their images are. The point of constructive empiricism is not lost if the line is drawn in a somewhat different way from the way I draw it.

1.2 Theories (and Theory-Ladenness)

Before discussing the notion that our theories can change so as to increase (or more generally alter) the scope of what we can observe, I should like to note that the admission that observations are theory-laden is double-edged for the realist. For example, one reason for which we may doubt the confirmation values ascribed to contemporary scientific theories, adduced in Chapter 3, is that observations are theory-laden. We want our theories to fit our observations. But if new theories can radically change the observations, in addition to the observation possibility space, then this tends to raise the probability that there are serious alternatives to existing theories. (One way of understanding this, recall, is that the evidence to be accounted for changes depending on the theories we have.) Moreover, to admit that what's observable is limited by what's theorised is to admit that there may be entities that are unobservable in principle, simply because the theories required to observe them are inconceivable. And this tends to raise doubts about the notion that we might find, or even get very close to, the truth.[8]

Another consequence of the thesis that what we can observe depends on our theories is that observables (in principle) may become *unobservable* (in practice) when theory change occurs.[9] Most obviously, this is in the case of things that 'exist' as a matter of convention, e.g. things that are discussed, in biology, as the result of taxonomical conventions. (These may be kinds, as well as instances of those kinds, *inter alia*.) A realist might respond that she is concerned merely, or failing that primarily, with our ability to observe instances of natural kinds improving as a result of theory change. But to this, granting that such kinds exist, I'd say two things.

First—in line with the discussion of models in Chapters 1, 2, and 4—theories (and thus observations) founded on conventions may often be better for practical purposes, including saving and understanding the phenomena, than those founded on natural kinds. More carefully, it should be allowed that admixtures of theories involving natural kinds and theories involving conventions might be optimal, or even the only satisfactory theoretical frameworks (in many contexts of inquiry). Second, even granting that we'd be able to observe every natural thing (in practice) if we possessed all true theories, and only true theories, it doesn't follow that increasing our stock of true theories, let alone increasing the truthlikeness of our theories, will typically tend to improve our ability to observe natural things (*qua* instances of natural kinds). Sometimes, indeed, what we can observe may be increased in scope by introducing a new false theory in a way it would not be by introducing a new true or approximately true theory (in the same domain).

Furthermore, one might distinguish between things that are significantly different for 'natural' reasons, without knowing of those reasons (and hence the relevant theories). For one may observe something while mistakenly thinking it is something else, in so far as one may see something without *seeing that* it is what it is. Consider 'planets' in medieval, pre-Copernican, astronomy. These were understood to be wandering stars—'πλανήτης' is the Greek for 'wanderer'. And Galileo observed Neptune on two separate occasions—28 December 1612 and 28 January 1613—without realising that it was a planet, as noted by Kowal and Drake (1980).

Thus the theory-ladenness of observations tells against scientific realism to a considerable extent. But perhaps it nevertheless renders the distinction between the observable and the unobservable less significant than many anti-realists take it to be? To examine whether this is so, let's consider two specific cases where Maxwell (1962: 14) alleges that direct observation of previously unobservable things can occur:

> [U]sing our painfully acquired theoretical knowledge of the world, we come to see that we 'directly observe' many kinds of so-called theoretical things. After listening to a dull speech while sitting on a hard bench, we begin to become poignantly aware of the presence of a considerably strong gravitational field, and as Professor Feyerabend is fond of pointing out, if we were carrying a heavy briefcase in a changing gravitational field, we could observe the changes of the $G_{\mu\nu}$ of the metric tensor.

My response has two parts. First, what one experiences in each situation may also be expressed in terms of (highly stable) *folk*-theory-laden observation statements. One might say "My bottom is hurting because of the pressure exerted on it by the bench" in the first case. That is, irrespective of whether one has the additional theories that the force acting on one's posterior is reactive (in line with Newton's third law), that one is pulled down onto the bench by gravity, and so forth. In the second case, one might say things such as "My briefcase is feeling heavier and lighter", or even "My briefcase is pulling more and less on my arm" (and even spot a pattern in the variation over time). As Feyerabend (1958: 163) elsewhere notes: 'the scientist . . . does not introduce a new use for familiar words such as "pointer", "red", "moving" etc. whenever he changes his theories'. Moreover, the truth of the aforementioned observation statements doesn't depend on, although it is clearly consistent with, the more technical statements that can be made by employing notions from physics. For instance, 'pressure' need not be understood in standard physical terms—i.e., as a force per unit area—by the speaker in the first example. The same goes for 'heavier', 'lighter', and 'pulling' in the second. Folk-theory-laden observation statements are more 'basic' than those employing scientific notions partly because one role of the latter is to save the former. That is to say, *'to save the phenomena'* may be understood as *'to save the phenomena on a proper subset of folk theories'*.[10] To illustrate, consider the way that iron filings 'arrange themselves' into distinct lines upon being scattered around a magnet. Is to see the iron filings' behaviour to observe a magnetic field? No, one might reasonably insist. Clearly, the folk-theory-laden statements about the situation—which make appeal only to 'iron filings', 'iron bar', statements about their relative positions and so on—might be true even if there were no magnetic field present.

Second, the statement that Maxwell attributes to Feyerabend is false even on the assumption that it is sometimes possible to observe changes 'of the $G_{\mu\nu}$ of the metric tensor'. That's because we don't always notice changing gravitational fields, according to physical theory. For example, this entails that the gravitational field on Earth is non-uniform. But I don't notice my backpack getting lighter when I walk up a mountain. Somewhat unfortunate for Maxwell's argument, moreover, is that whether the briefcase is 'heavy' depends, on a normal physical interpretation, on the field strength. So to make sense of the prose, one has to read 'heavy' in a non-physical, folk, sense. (This also raises the interesting topic of mass. How would one maintain that we observe this, as distinct from observing inertial mass and gravitational mass? Isn't it typically *inferred* that there's one property, mass, because the aforementioned appear to have the same values?)

In summary, the distinction between 'observation' and 'detection' is vague, but 'detection' cases stereotypically involve long and explicit chains of inference. 'Observation' cases stereotypically don't involve any explicit chains of inference, and are laden on highly stable folk theories.

2. Comparison With Other Positions in the Realism Debate

In the final part of this chapter, I'll explain how cognitive instrumentalism compares with other alternatives to scientific realism. I'll also offer critiques of these alternative views.

I'll cover structural realism, constructive empiricism, and semirealism. I don't cover some other alternatives because it should be clear how the position I've developed differs from, and involves a principled rejection of, these. For example, I have argued against Hacking's variant of entity realism, according to which '*if you can spray them then they are real*' (Hacking 1983: 23), to some extent already. In particular, I have argued, following Resnik (1994: 395), that 'experimentation is not nearly as theory-free as Hacking maintains'. I take it to follow that 'the experimental realist can only have knowledge about theoretical entities if she assumes that the theories which describe those entities are at least approximately true' (Resnik 1994: 395). More precisely—on a topic to which I will return in the discussion of semirealism—I hold that one typically needs to appeal to scientific theories to infer the presence of theoretical entities in experimental contexts. To illustrate, here's one of Hacking's claims about electrons:

> Uhlenbeck and Goudsmit in 1925 assign angular momentum to electrons, brilliantly solving a lot of problems. Electrons have spin, ever after. The clincher is when we can put a spin on the electrons, polarize them and get them thereby to scatter in slightly different proportions.
>
> (Hacking 1983: 274)

As we saw in Chapter 4, however, the history is much more complex. Moreover, there are different theoretical stories about what causes the outcomes in the class of experiments that Hacking mentions.[11] And as we've seen, not all of those theoretical stories involve imputing intrinsic angular momentum to electrons, or even thinking that 'spin' refers to an intrinsic property of electrons. Bohm's theory is a case in point.

Indeed as Gelfert (2003) points out, quasi-particles, such as electron holes, can be manipulated despite not existing (from a common sense, as well as a contemporary theoretical, perspective). And 'if you don't know what you are spraying, you cannot tell whether it is real or not' (Gelfert 2003: 261). Even if one is spraying real things, moreover, it doesn't follow that one is spraying only one *kind* of thing.

I don't cover other alternatives to scientific realism because they haven't yet attracted much attention. Consider, for example, the 'singularist semirealism' advocated by Nanay (2013), according to which singular scientific statements should be construed in a realist fashion, whereas non-singular statements should be construed in an anti-realist fashion: 'there is always a fact of the matter about whether the singular statements science gives us are literally true, but there is no fact of the matter about whether the non-singular statements science gives us are literally true' (Nanay 2013: 371). All I will say is that in so far as this position is semantic in character, it's silent, strictly speaking, on issues such as the value of science (which I tackled in Chapter 1), and doesn't entail much about what science can be expected to achieve (which I tackled in Chapter 3). On the latter issue, for example, one might accept that all singular statements should be taken literally but deny that there are typically good grounds for thinking that such statements should be taken to be true, or even approximately true, when they concern unobservable things.[12]

2.1 Structural Realism

Structural realism comes in two flavours, epistemic and ontic. Put crudely, the former says 'Structure is all we can know concerning the unobservable aspects of the world', whereas the latter says 'Structure is all there is'. Or to use the words of influential structural realists, epistemic structural realism holds 'the objective world is composed of unobservable objects between which certain properties and relations obtain; but we can only *know* the properties and relations of these properties and relations, that is the *structure* of the objective world' (Ladyman 1998: 412), whereas ontic structural realism holds 'structure is ontologically basic' (French and Ladyman 2003a: 46). Naturally, I don't intend to tie advocates of these positions to such relatively crude characterisations. For example, an epistemic structural realist might want to say instead, more carefully, 'Well-confirmed theories in mature sciences approximately capture the structure of the unobservable parts of the world in their domain'. And ontological structural realism involves several variants too, as outlined by Ainsworth (2010). To put it simply, each variant involves holding that relations are ontologically primitive, while denying that objects or properties are. The use of 'or' is inclusive. (The most influential variant, associated with French and Ladyman, is logically the strongest: 'relations are ontologically primitive but objects and properties are not' [Ainsworth 2010: 51]. A weaker or more 'moderate' variant, where only properties aren't considered primitive, is defended by Esfeld and Lam [2008].) Thus, I will explain how my position differs from these approaches, and why I think this is a good thing, while understanding them, as I have scientific realism, in a relatively broad and permissive fashion.

I will deal with ontic structural realism first. Cognitive instrumentalism allows that discourse concerning unobservable entities—objects and properties, not merely relations—may be taken literally. It also allows that some of our discourse concerning unobservable entities is true (although it adds that we lack reliable means to determine which parts of said discourse are true). Thus, it allows that unobservable entities falling into those categories exist. Moreover, as I've already mentioned, I think that (but need not think, *qua* cognitive instrumentalist, that) they probably do. I agree with Maxwell (1962), as explained earlier, that the distinction between observable and unobservable has no ontological significance.

I will explain why I object to ontic structural realism in short order. But I would like to preface this by saying something positive about the motivation behind the view. Authors such as French and Ladyman (2003a, 2003b)—see also Ladyman and Ross (2007)—look at contemporary, or at least very recent, physics (concerning the microworld, or more accurately the sub-microscopic world); and they use this as the basis for their ontological claims. This is praiseworthy in an important respect; they strive to understand and engage with areas of fundamental physics that some philosophers of science do not. Moreover, their findings about that physics—e.g. what it says or intimates about identity and individuality (French 1989)—are interesting and valuable in their own right.

My primary concern with ontic structural realism, however, is that it's too epistemically presumptuous. Even assuming that 'structure is all there is' follows from modern science, that's to say, why should we think that this will follow, or is likely to follow, from future science?[13] I think that the structural realist claims to know more than she does. Instead, I think intellectual humility is appropriate. To put matters bluntly, it often seems reasonable to take discourse about unobservable objects and their properties literally, as I argued in Chapter 2. I thus take ontic structural realists to claim to have determined that all such discourse (about the objects and/or about their properties) is false. I trust that this way of putting things makes it clear how strong the claim of ontic structural realists is. All that underpins it, as far as I can determine, is trust in part of contemporary science. And as I argued in Chapter 3, there are reasonable grounds to hold that this trust is misplaced.

Ontic structural realism is also a narrower position than cognitive instrumentalism in so far as it's relatively silent, strictly speaking, on the value of science.[14] For instance, I have argued that increasing understanding is a central aspect of scientific progress, and that this is often achieved by talk about everyday objects and processes (even when this should be taken to be non-literal). Therefore, I doubt that talk about objects is eliminable from (good) science. And while this is consistent with ontic structural realism, it's a matter on which the position is silent. It might therefore be said that ontic structural realism doesn't involve a firm position on the axiological or semantic aspects of the realism debate. Hence a structural realist one could agree with cognitive instrumentalism to a considerable extent. She could agree, for instance, with the core theses advanced in Chapters 1 and 2.

Epistemic structural realism is also narrower than the instrumentalism defended herein, when it comes to what it says about—and what it implies about—the value of science (or scientific progress).[15] It also differs in other key ways. For instance, Worrall (1989b: 122) states that: 'the structural realist . . . insists that it is a mistake to think that we can ever "understand" the *nature* of the basic furniture of the universe.' As explained in Chapter 2, however, a property instrumentalist might hold that we're able to understand the nature of such basic furniture to the extent that it involves observable properties. (I take Worrall to use 'understanding' in a factive sense.) Indeed, it's possible that some such items of furniture might be fully comprehensible. But to understand something—in the sense of having true beliefs about that thing and all its properties—need not be to have strong evidence *that* one understands it; so, in line with what's argued in Chapter 3, we might say that we're typically unaware of the extent to which we understand unobservable entities (and as a result, observable entities composed of unobservable entities).

As intimated earlier, Worrall also appears to overlook the possibility of construing 'understanding' in non-factive terms, and perhaps, as a result, the importance of trying to generate models that are comprehensible in terms of the everyday world of experience. He writes:

> [T]he structural realist simply asserts . . . that, in view of the theory's enormous empirical success, the structure of the universe is (probably) something like quantum mechanical. It is a mistake to think that we need to understand the nature of the quantum state at all; and *a fortiori* a mistake to think that we need to understand it in classical terms.
>
> (Worrall 1989b: 123)

I agree that it's a mistake to think that we need to have a true account of the unobservable world (and hence 'the quantum state'). But an important task of science, on the view I defended in Chapters 1 and 2, is to tell a story in terms of observables when it's possible. So understanding 'in classical terms' is important, to the extent that it's possible, in line with the approach exemplified by many of the scientists discussed in Chapter 4.

One final comparative point is worth making. A difficulty for the structural realist is: what should we believe in now, in our contemporary theories? As van Fraassen (2006: 290) points out:

> [T]he division between structure and content seems never discernible beforehand. (The structure discovered is identifiable only in retrospect—it is the part retained through scientific theory change. . . . The atoms are still there at some level, so that was structure. The ether is no longer there, at any level, so that was a mistake about content. . .). If that is so, we have to wonder whether there can then be an independent criterion for this division at all.

Even granting there is an independent criterion for division, moreover, there's the question of which parts of the structure are operative, and which parts are not. I call this the problem of excess structure. Consider, as an illustration, the following equation for the total energy of a massive particle in special relativity:

$$E = \frac{m_0 c^2}{\sqrt{1 - \dfrac{v^2}{c^2}}}$$

This equation has the desirable feature that the energy of a massive particle would have to be infinite for it to reach the speed of light; naturally, this is taken to mean that it's not possible for such particles to reach the speed of light. There is, one might say, a speed of light *barrier*. However, this is compatible with allowing rest masses to take imaginary values, and the existence of particles constrained to move faster than the speed of light; see Bilaniuk et al. (1962). (That is to say, if one assumes that energy must be real valued. If one drops that assumption, one could have massive particles moving faster than the speed of light with imaginary energies.) The point here is that even when the equation is interpreted to a certain extent—m is taken to represent an object's mass, v its velocity, E its energy, and c the speed of light—there is arguably more structure than is needed. I suppose an epistemic structural realist, such as Worrall, would say that m being real valued should be part of the package—that we should consider only the structure when this constraint is met. However, this is not so clear as it may first appear. As Bilaniuk et al. (1962: 720) explain:

> [I]maginary 'rest mass' . . . may seem to disqualify the whole idea right from the start. One should recall, however, that in classical mechanics the mass . . . is a parameter which *cannot* be measured directly even for slow particles. . . . Only energy and momentum, by virtue of their conservation in interactions, are measured, therefore must be real. Thus the imaginary result for the rest mass of the hypothetical "meta" particles offends only the traditional way of thinking, *and not observable physics*. [emphasis added]

This issue of excess structure is not significant if one is concerned merely with (a realistic interpretation of) structure at the level of the phenomena.[16] As van Fraassen (2006: 303) notes:

> [W]e can . . . prevent ourselves from sinking into [the] metaphysical morass that swallows all seekers for the true foundations of being. The empirical successes of the older theories were partial successes of a very distinct sort: their representations of nature, the models they made available for representation of the observed phenomena, were partially accurate. These successes consisted in their success of fitting

the data, the deliverances of experimental and observational experience. *There was something they got right: the structure, at some level of approximation, of those phenomena.* [emphasis added]

Indeed, cognitive instrumentalism is compatible with the view that we can reasonably expect science to discover more about such *empirical* structures, and indeed that science progresses, in part but not in whole, via making such discoveries. In short, that's to say, it's compatible with some possible variants of structural empiricism (such as the version defended by Bueno 1999; 2011).[17]

2.2 Constructive Empiricism

As I have already referred to constructive empiricism at several points in the book, most notably in Chapter 2, it has already emerged how it differs from cognitive instrumentalism in some respects. Most obviously, cognitive instrumentalism links meaning to observability, and hence involves taking some scientific discourse non-literally, in a way that constructive empiricism does not. As will become clear later, however, a constructive empiricist *might* accept this (property instrumentalist) view on meaning, although van Fraassen rejects it.

Constructive empiricism is defined as follows:

> Science aims to give us theories which are empirically adequate; and acceptance of a theory involves as belief only that it is empirically adequate.
>
> (van Fraassen 1980: 12)

Two initial remarks about this definition are in order. First, it involves vague and misleading talk of the aim of science. At best, this requires laborious unpacking. But since I address this problem in the Appendix, I shall say no more about it here other than that van Fraassen elsewhere describes 'the aim of science' as 'what counts as success in science' and denies that this should be understood sociologically.

Second, van Fraassen (1980) uses this definition for contrastive purposes *because* he defines scientific realism in a similar way, namely as a thesis about aims and acceptance. But this is a mischaracterisation of scientific realism, for reasons I also explain in the Appendix. For instance, scientific realists are typically committed to an epistemic descriptive claim such as 'Well-confirmed theories in mature science are typically approximately true'.[18] Constructive empiricism says nothing about this. As van Fraassen (2007: 343) puts it—here, taking himself as an example of a constructive empiricist—'I do *not advocate* agnosticism about the unobservable, but claim that belief is supererogatory as far as science is concerned.'

Now let's consider acceptance, which van Fraassen understands to involve belief and a measure of commitment (to use the theory for various purposes). As Monton and Mohler (2014) explain:

> One reason the constructive empiricist's account of acceptance is important is that it allows us to make sense of scientific anti-realists such as constructive empiricists (of the scientific agnostic variety) talking as if a particular theory is true. When one looks at scientific discourse, this is what scientists are often doing: they treat a theory as if they fully believe it, and answer questions and give explanations using the resources of the theory. The constructive empiricist can account for this behavior, without attributing full belief in the theory to the scientists, by describing the scientists as merely accepting, without fully believing, the theories they develop.

My own view, by contrast, is that this stratagem is unnecessary. (And I don't want to burden scientific realists with a view on acceptance. I don't think this is fair, even if *some* realists take a view on it.) There are many examples from the history of science—I presented several in Chapters 2 and 4—of scientists treating theories as if they fully believed them in some contexts, while explicitly denying that they fully believed them in others.

Moreover, the appeal to empirical adequacy as 'what counts as success in science', on the view I developed in Chapter 1, is both too narrow and too broad. (Although van Fraassen allows for 'subsidiary' aims, so the dispute here is really over what should be considered central to 'success', or, to use the term I prefer, 'progress'.) It's too narrow in so far as it doesn't acknowledge the significance of fostering understanding.[19] It's too broad in so far as progress can be made in science—and 'success' can plausibly be achieved—when rather less than empirical adequacy (of the content) is achieved. To be specific, it's typically acceptable if what a theory (or model) says about the phenomena is wrong to some extent. For one thing, it could be *approximately* true in what it says about the phenomena (in conjunction with appropriate statements of initial conditions, or auxiliary statements). It could be 'true enough' for our purposes in a range of significant contexts, as classical mechanics is in many situations. For another, what a theory (or model) says about the phenomena in some situations might be true, whereas what it says about the phenomena in others might be false (again, in conjunction with appropriate auxiliary statements). And this is fine as long as we only use the theory (or model) in the appropriate situations. As I argued in Chapters 2 and 4, this kind of piecemeal approach fits anti-realism more naturally than it fits scientific realism.[20]

Finally, there's a potential problem with van Fraassen's personal view on science—albeit not constructive empiricism, strictly speaking—that cognitive instrumentalism avoids. Ladyman (1998: 417 & n. 11) puts it so:

> Although agnostic about scientific realism, he is atheistic about modality and believes in no objective causality or necessity in reality. . . . There is therefore a tension in van Fraassen's position, for he thinks that unobservable entities *may* exist, but that laws and natural kinds definitely do not. Yet if, for example, electrons do indeed exist but do not form a natural kind subject to laws of interaction and so on, then we are wrong about a very important aspect of them, making it hard to see on what grounds we can say that *they* exist rather than some other things entirely.

Cognitive instrumentalism is instead neutral on, and compatible with belief in, the existence of natural kinds and (either resultant or independent) laws of nature. In effect, the existence of such things is seen as irrelevant; for while the presence of regularities would be guaranteed by their existence, it would not be ruled out in virtue of their absence. And regularities provide the basis for saving the phenomena (and understanding how they interrelate), even if only in a piecemeal fashion.

2.3 *Semirealism*

Central to semirealism is the distinction between two kinds of properties, namely detection properties and auxiliary properties. In the words of Chakravartty (1998: 394–395):

> We infer entity existence on the basis of perceptions grounded upon certain causal regularities having to do with interactions between objects. Let us thus define *detection properties* as those upon which the causal regularities of our detections depend, or in virtue of which these regularities are manifested. *Auxiliary properties*, then, are those associated with the object under consideration, but not essential (in the sense that we do not appeal to them) in establishing existence claims. Attributions of auxiliary properties function to supplement our descriptions, helping to fill out our conceptual pictures of objects under investigation. Theories enumerate both detection and auxiliary properties of entities, but only the former are tied to perceptual experience.[21]

Semirealism involves the view that detection properties should be 'epistemically up valued'—I use this phrase to avoid becoming embroiled in a debate about the difference between belief and commitment—in a way that auxiliary properties should not be. As Chakravartty (1998: 404–405) explains:

> Semirealism is a commitment to those detection properties of entities which underwrite our detections and compose structural relations. The limited epistemic vision of this position incorporates auxiliary properties, not as substantive knowledge, but as methodological catalysts . . . the task of inquiry is to attempt either to falsify them, or to transform them into detection properties in virtue of which previously unknown relations are brought to light.

It therefore becomes clear that there are three key ways in which semirealism differs from cognitive instrumentalism. First, despite its shared emphasis on properties, semirealism doesn't say anything explicit about when talk of auxiliary properties should—and more precisely, should not—be construed literally. Indeed, Chakravartty (1998: 403) advocates a version of semirealism according to which: 'descriptions of entities which are inclusive of all properties attributed to them within the confines of some theory have meaning'. Second, semirealism involves a different stance on what's central to scientific progress; that is, provided 'the task of inquiry' can be appropriately related to what's central to making progress. Detecting new properties takes centre stage, as opposed to 'improving predictive power and understanding of how phenomena interrelate'. Third, semirealism is more sanguine about the prospects of science discovering new truths about unobservable things (i.e., by converting some auxiliary properties into detection properties).

In closing, I'll also offer a brief critique of semirealism. Let's do the semirealist the favour of granting whichever view of causation she prefers. (For instance, Chakravartty [2005] prefers 'causal realism', which involves positing *de re* necessity.[22] Cognitive instrumentalism is neutral on such issues.) Let's also grant that science sometimes progresses in the way Chakravartty outlines. That's to say, let's grant that we sometimes detect unobservable things via (detecting or observing) a proper subset of their properties, and make ('auxiliary') posits about further of their properties not yet detected which go on to be detected—to become 'detection properties'—themselves. I nevertheless think it's implausible that we can typically tell when we're doing this (or have done this) successfully, and when not, for reasons I'll explain later.

One simple concern is as follows. How do we know when we're dealing with a detection property, and when we're dealing with an auxiliary property? Chakravartty (1998: 394) initially describes 'detection properties' as follows:

> [D]efine *detection properties* as those upon which the causal regularities of our detections depend, or in virtue of which these regularities are manifested.

Later in the same paper (402), however, he remarks that:

> [D]etection properties [are those] on the basis of which we infer entity existence.

It seems, however, that the properties upon which the causal regularities of our detections depend (or in virtue of which said regularities are manifested) may *not* be those on the basis of which we infer entity existence.[23] (A problem here is that the definition is rather vague. Metaphysically speaking, it's easy to see that causal regularities might depend on numerous properties, observable and unobservable. It is hard to see how we would recognise the presence of a property merely because it was responsible for a causal regularity).[24]

Even granting that we know we have a detection property, moreover, we may be wrong about to which putative entity, or group of entities, it should properly be ascribed. Consider, for example, the ingenious experiments performed by Berzelius in order to determine the (relative) atomic mass (or weight) of chlorine, which he described as follows:

> I established its [chlorine's] atomic weight by the following experiments: (1) From the dry distillation of 100 parts of anhydrous potassium chlorate, 38.15 parts of oxygen are given off and 60.85 parts of potassium chloride remain behind. . . . (2) From 100 parts of potassium chloride 192.4 parts of silver chloride can be obtained. (3) From 100 parts of silver 132.175 parts of silver chloride can be obtained. If we assume that chloric acid is composed of 2 Cl and 5 O, then according to these data 1 atom of chlorine is 221.36. If we calculate from the density obtained by Lussac, the chlorine atom is 220 [relative to the atomic weight of oxygen]. If it is calculated on the basis of hydrogen then it is 17.735.
>
> (Quoted in Harré 1981: 206)

The final value mentioned in the quotation is very close to half of what it should be, according to the contemporary periodic table. However, the fact that it is a factor of two off is explained, we would say on the basis of contemporary science, by the fact that hydrogen gas is diatomic rather than by any deficiency of Berzelius's experiments (1)–(3). Thus, we may say that Berzelius *mistook* a property of a molecule for a property of an atom. And it seems hard for a semirealist to deny that said property was a detection property, as opposed to an auxiliary property, given the agreement between Berzelius's value and the contemporary value.

Moreover, Berzelius quite reasonably took himself to have discovered the atomic mass of *each and every chlorine atom*, although a semirealist would have to concede, on the basis of modern science, that he instead found a property of *a collection of different types of chlorine atom*. Or to put it more tersely, we now take the detection property to belong to different isotopes of chlorine in a specific ratio. Moreover, we think this ratio can vary in any given sample, and hence shouldn't be taken to correspond to a property of any natural kind of thing. (The case that we're dealing with a detection property is, again, very strong. That is, by the semirealist's own lights. The experiments were performed with the specific aim of detecting the kind of property that they did, and modern science tells us that they did detect a property, albeit of a mixture.) More generally, we may ask the following. How do we know which clusters of detection properties belong together? How do we know which are detection properties of the same thing (or group of things, or what have you)?

One might respond that we can use our theories to help us to determine what our detection properties are detection properties of (and/or how detection properties should be bundled). But to do so is to retreat from semirealism, back towards scientific realism; that is, in so far as the approximate truth of those theories would presumably be required, at the bare minimum, to make this a reliable process. (More precisely, it's dubious that one could typically achieve the task without appeal to any auxiliary properties.) It therefore seems that semirealism has to be weakened, if it's to maintain its distinctive character, to the view that we sometimes realise (or correctly believe that) we're detecting properties of things, although we typically can't be reasonably sure how many things we're detecting the properties of, or even whether those things are natural kinds. This seems closer to anti-realism than realism. (It might even collapse into a version of epistemic structural realism.) It is compatible with cognitive instrumentalism.

3. Conclusion

In this chapter, I've done two things. First, I've defended the significance of the distinction between the observable and the unobservable (with respect to the theses argued in previous chapters). Second, I've compared cognitive instrumentalism with other key alternatives, and explained why I take it to be preferable.

In closing, I'd like to reemphasise two things. First, cognitive instrumentalism is modular, and the parts are defensible in isolation (although they cohere). For example, you could be persuaded by my arguments on the value of science and scientific progress, in Chapter 1, without being persuaded by the property instrumentalism developed in Chapter 2 or the extension of the argument from unconceived alternatives in Chapter 3. And so on, *mutatis mutandis*.

Second, cognitive instrumentalism allows for a more moderate view on the observable-unobservable distinction than the one I've argued for. For instance, one might agree with each of its component theses, yet think that most of what's unobservable in practice *will* become observable in practice. One would then be a cognitive instrumentalist who is much more optimistic about the prospects of future science than I am.

Notes

1 By way of contrast, as van Fraassen (1985: 258) notes, 'scientific realists tend to feel baffled by the idea that our opinion about the limits of perception should play a role in arriving at our epistemic attitudes toward science'.
2 See Bradie (1986) on the EEM ('Evolution of Epistemological Mechanisms') programme.
3 See Okasha (2006) for an extensive treatment selection at different levels.
4 Clearly we may legitimately question the results of some of our observations, e.g. concerning the relative length of lines in the Müller-Lyer illusion. However, such questioning is on the basis of other observations using our senses, such as measurements with rulers. That's to say, 'being stuck with' using our instruments is compatible with preferring their 'outputs' in some circumstances over those in others. It's worth adding that the relevant 'output' in this case—the way the lines appear—doesn't, plausibly, depend merely on the eye. See, for instance, Segall et al. (1963). However, the basic point stands even when the relevant 'instrument' is individuated in a more sophisticated way.
5 Hacking (1981: 138) adds: 'considering the optical aberrations it is amazing that anyone ever did see anything through a compound microscope'.
6 SNO stands for 'Sudbury Neutrino Observatory' and LIGO stands for 'Laser Interferometer Gravitational-Wave Observatory'.
7 Van Fraassen (2001: 159–160) provides a useful taxonomy of images:

> [I]n the case of images I want to describe a division into several kinds and subkinds.
>
> *Type 1.* On one side are the images which are in fact things, such as paintings and photos.
> *Type 2.* On the other extreme are the purely subjective ones like after-images and private hallucinations. These are personal, not shared, not publicly accessible. Indeed, we are pretty clearly dealing there with discourse that reifies certain experiences which are "as if" one is seeing or hearing.
> *Type 3.* In between these two are a whole gallery of images which are not things, but are also not purely subjective, because they can be captured on photographs: reflections in the water, mirror images, mirages, rainbows. For those I will use the term 'public hallucinations'.

8 One might respond that the possible is a subset of the conceivable. Suffice it to say for present purposes, however, that it's reasonable to doubt this without a strong argument to the contrary. The claim is exceptionally bold in the light of psychological evidence about human reasoning deficiencies and the limits of 'creative thought'. On the latter, see Ward et al. (1997).
9 Compare the idea of 'Kuhn loss', and note that one can't occupy two theoretical frameworks at the same time, even if one can swap between them (in a way analogous to Gestalt switching).
10 We need the 'proper subset' qualification because some folk theories do have to be abandoned in the face of scientific theories. Said theories are typically not used to express individual experiences. The proper subset may be characterised precisely by its stability.

11 There are other points where I strongly disagree with Hacking (1983) on the history of science. For example, he claims 'In the early 1920s an experiment by O. Stern and W. Gerlach suggested that electrons have angular momentum' (Hacking 1983: 84). I explained why this is wrong in Chapter 4.

12 My view is that the *content* of the singular statements, not merely their form, is important in deciding whether they may be construed literally. Ditto for non-singular statements. This view is intimated by the discussion in Chapters 2 and 4.

13 I doubt that this statement does follow from science, moreover, unless one accepts a reductionist line. That is to say, even if it's consistent with one key branch of physics. As suggested in my discussion of folk-theory and observation, I take the objects of everyday experience to be as much a topic of contemporary science as their fundamental parts are.

14 Clearly the advocates of ontic structural realism take positions on this. But these aren't integral parts of ontic structural realism.

15 In fact, epistemic structural realism has direct and indirect variants, as explained by Frigg and Votsis (2011: section 3). These differ in so far as they draw the distinction between the observable and the unobservable differently. Roughly, indirect structural realists take us not to be acquainted (in Russell's sense) with physical objects, as opposed to sense data (or something similar), and therefore classify physical objects as unobservable. As such, they hold that we should not expect science to arrive at approximate truths regarding any such objects. Direct structural realists disagree, and take us to be acquainted with many physical things. Cognitive instrumentalism is compatible with either view on what's observable.

16 Van Fraassen (2006: 297) points out a further problem with structural realism, namely that it is somewhat metaphysically presumptuous:

> [Worrall's] claim requires a context-independent, 'objective' division in nature between mere quality and the relational structure in which the qualities appear at the vertices, so to speak. This can be made good only by falling back on a (here unacknowledged) metaphysics in which such an intrinsic-extrinsic distinction makes sense. I rather doubt that today's structural realists in philosophy of science are very anxious to creep into that thicket . . . but without that, and without a way of finessing the point, the burden of unacknowledged metaphysics rests heavily on their position.

17 Bueno (1999) describes structural empiricism as an extension to constructive empiricism. However, it might also serve as an extension to, or even a complement to, cognitive instrumentalism and other such anti-realist positions. It might require some reformulation. For example, Bueno (1999, 2011) cashes out structural realism with extensive reference to the notion of empirical adequacy. Nonetheless, a cognitive instrumentalist might agree that 'Structures are employed not to capture the "structure of the world"; they only help us represent what is observable' (Bueno 2011: 96), and that 'there is a [significant] distinction between *ontological commitment* and *quantifier commitment*' (Bueno 2011: 97). Indeed, I agree wholeheartedly with the latter. I agree with the spirit of the former, although would say, instead, something like: 'The structures used in science rarely, if ever, capture "the structure of the world", although they often capture relationships between observable things' or 'We can't tell if any of the structures used in science capture "the structure of the world", although we can tell that some structures used in science fail to capture "the structure of the world" and that some capture (some of) the structure of the phenomena'.

18 Indeed, van Fraassen (2006: 288) later concedes this, it seems, when he instead characterises scientific realism as follows:

> The aim of science is to provide us with a literally true story about what there is in the world, *and this aim is actually achieved to a great extent, because it is pursued by effective means to serve that end.* [emphasis added]

19 It also doesn't explicitly acknowledge the significance of enabling interventions, and fostering know how, as I did in Chapter 1. As I explained there, however, a charitable reading of 'saving the phenomena', and hence 'achieving empirical adequacy', covers this.

20 Van Fraassen likely agrees with much of what I say in this paragraph; but then he should change the definition of constructive empiricism.

21 See also Chakravartty (1998: 402):

> By 'entities' we refer to causal substances, which we associate with two kinds of properties: detection properties, on the basis of which we infer entity existence, and auxiliary properties, which further describe or supplement our conception of the entities present within a particular theory.

22 See also Chakravartty (2007: 91–92).

23 Psillos (2013: 36–37) raises some closely related concerns:

> [The] distinction between detection properties and auxiliary properties is a central plank of semirealism. I am not sure, however, it is carefully delineated. It is clearly meant to be an epistemic distinction—one that is related to our state of knowledge, that is to what we already know by having causally interacted with. Chakravartty claims that "Detection properties are causal properties one has managed to detect; they are causally linked to the regular behaviours of our detectors. Auxiliary properties are any other putative properties attributed to particulars by theories" (p. 47). This distinction, however, is moveable—some auxiliary properties may be 'converted into detection properties'; others may be simply jettisoned.
>
> So the distinction seems to be more pragmatic than epistemic. There is no epistemic mark of being auxiliary apart from the fact that there has not as yet been a causal detection of the property that is characterised as auxiliary. But causal detectability is always a matter of degree, unless a property is either causally isolated or inert. Detection can be more or less direct. Most properties are detectable by long causal chains of actions and interactions and there is no clear and sharp distinction between being detectable and being undetected (unless, as noted already, a property is already taken to be causally inert or isolated). [Internal reference is to Chakravartty (2007)]

Part of Chakravartty's (2013) reply, which is reasonable, is as follows. One's confidence that a given property is a detection property can, and should, vary depending on the available evidence. However, this doesn't help with the objection I raise later.

24 I assume Chakravartty had something else in mind than his definition read literally, but I find it difficult to discern what. My own way of talking about the use of instruments, earlier, is to allow that sometimes properties that were unobservable can become observable. But I take 'detection' to involve the use of long inferential chains that aren't present when observation occurs. This is in line with my earlier discussion of images, and the next point in the main body.

7 The Illusion of Scientific Realism

Scientific realism has a curious attraction, which can persist even when effective arguments against it are enumerated. So why does this pull exist? My task in this final chapter is to argue that this is because our intuitions are—or 'common sense' is—apt to deceive us on this matter.

More carefully, I will argue for two claims that I take to be supported by experimental psychology. First, people are overconfident about what they know when it comes to causal explanations. Second, people are especially overconfident about what they know when it comes to causal explanations of observable entities in terms of unobservable entities.

These combine to paint a dim picture of our normal estimations of science. We are not only overconfident in what science says about the world in general. We are supremely overconfident, in addition, of the explanations it gives in terms of unobservable things and mechanisms. It appears to us that we can (truthfully) explain how observable things behave and interrelate, in terms of the unobservable, when our understanding is 'surface'. This is the basic idea. But the full argument is more nuanced. For instance, expertise may diminish confidence levels to some extent, and hence make them somewhat more appropriate.

1. The Illusion of Explanatory Depth

Several studies underpin a widespread consensus in psychology that people are overconfident in general; that they are overconfident not only about their propositional knowledge, but also about their other competences.[1] Kruger and Dunning (1999), for example, performed three studies to examine how subjects rate their humour, logical reasoning, and grammatical ability in comparison to that of others.[2] Overestimation was present in all three studies. Moreover, those in the bottom quartile on the competence measures overestimated their ability by a significantly greater margin than others.

However, such general overconfidence would be of no special interest to the case we're considering, namely belief in scientific realism. Indeed, it might be used to argue that anti-realists are just as liable to be overconfident in their anti-realism as scientific realists are in their scientific realism. So it is worth highlighting in advance that the studies considered herein concern a distinct kind of overconfidence. In the words of Rozenblit and Keil (2002: 522):

> The illusion for explanatory knowledge—knowledge that involves complex causal patterns—is separate from, and additive with, people's general overconfidence about their knowledge and skills.

So why do Rozenblit and Keil say this? What did they do? In their first four studies, there were five distinct phases. In the first phase, which was instructional, subjects were introduced to a 7-point scale of understanding and shown how it should be used to differentiate accounts of: (a) how a crossbow works, and (b) how a GPS (global positioning system) works. Specifically, they were given examples of what would constitute level 1 (minimal) understanding, level 4 (intermediate) understanding, and level 7 (maximal) understanding of (a) and (b). These were in the form of diagrams with accompanying text.

In the second phase, the subjects were asked to rate their understanding of 48 items on the 7-point scale, without pausing too long on any particular item on the list. Then, in the third phase, the subjects were asked to provide written explanations of how four of the items mentioned previously operate. There were two groups of four items used, each for half of the subjects in total. Group one was a flush toilet, a piano key, a speedometer, and a zip. Group two was a cylinder lock, a helicopter, a quartz watch, and a sewing machine. After providing each explanation, the subjects again rated their understanding on the scale.

In the fourth phase, the subjects were asked to answer a question about each of the items, designed to probe their understanding: 'For example, they were asked to explain, step-by-step, how one could pick a cylinder lock.' (Rozenblit and Keil 2002: 528). After doing so, they were asked to rate their understanding on the scale again.

Finally, in the fifth phase, the subjects were provided with an expert explanation of how each item works, including text and diagrams. They then re-estimated their initial understanding in the light of this new information. They were also asked to rate their understanding at the end, as a result of their exposure to the expert explanation.

Figure 7.1 shows a diagram of the process and Figure 7.2 displays a graph showing the results from the first two studies performed (which are similar to those in the other two).

The primary feature of interest is the dip in the curve. When asked to apply or show what they allege to know (T2 and T3), the subjects' confidence in knowing it diminishes significantly. And when they are exposed to an expert display (T4), they maintain that their initial estimation of their knowledge was incorrect. Finally, their response to the expert

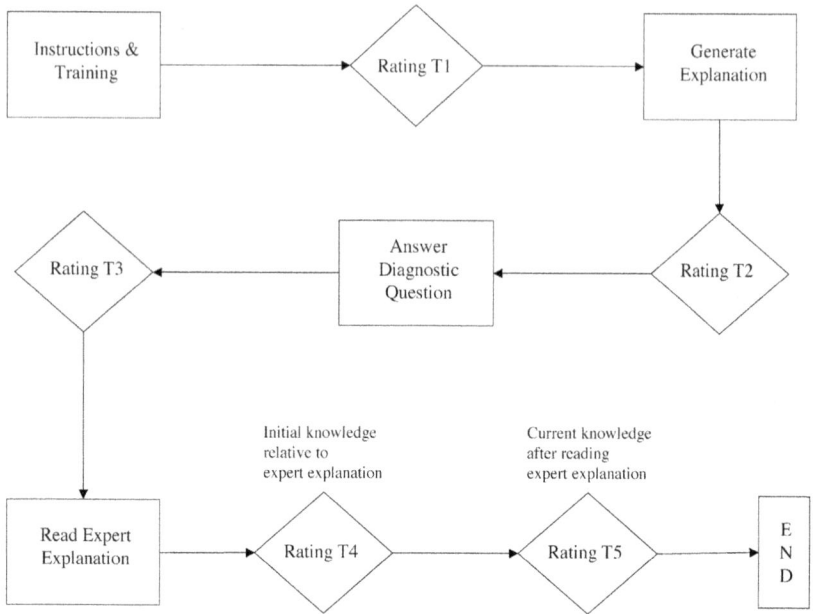

Figure 7.1 Procedure in Rozenblit and Keil's (2002) First Four Studies
Reproduced with permission from Rozenblit and Keil (2002). ©2002 Leonard Rozenblit.

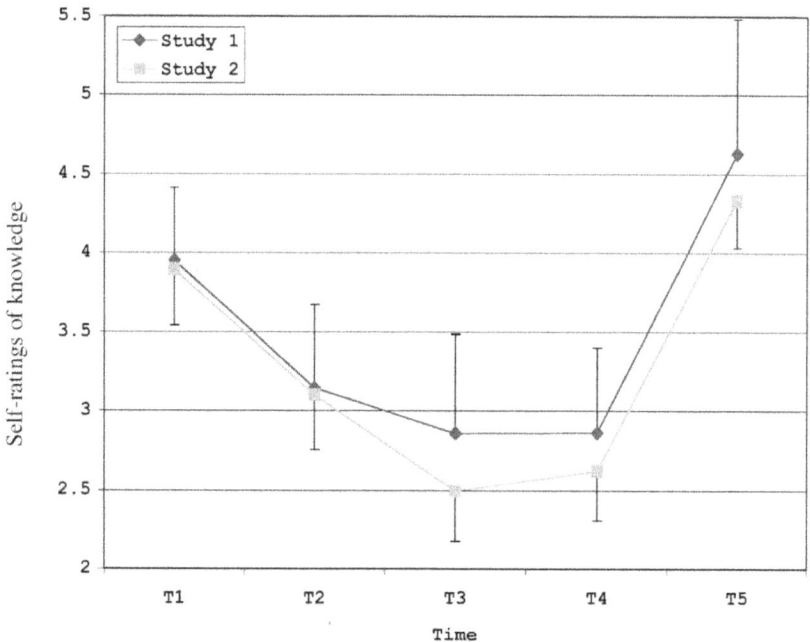

Figure 7.2 Results from Rozenblit and Keil's (2002) First Two Studies
Reproduced with permission from Rozenblit and Keil (2002). ©2002 Leonard Rozenblit.

display is to claim greater knowledge than they previously claimed (T5). Rozenblit and Keil (2002: 530) remark that many of the participants reported 'genuine surprise and new humility' about their lack of knowledge in the debriefing sessions, although several nevertheless maintained that they would have done better if they had been assigned the other set of four items to discuss! However, there was no significant difference in the performances of those using either group.

The responses to the final two tests, T4 and T5, serve to rule out two salient possibilities. The former indicate that the participants did not think that the diagnostic questions asked before T3 were unreasonably detailed, because, in that event, the values given at T4 would have increased: 'After all, if the experts did not think a piece of knowledge is critical why should the participant?' (Rozenblit and Keil 2002: 530). The results on the latter rating rule out the possibility that the experiment was shaking confidence continually.

The studies were also designed to rule out some other possible explanations of the effect noted. In the first study, the subjects were graduate students from Yale. Hence, by using undergraduates from Yale instead, the second study ruled out the amusing hypothesis that 'graduate study leads to an intellectual arrogance' (Rozenblit and Keil 2002: 530). Study three involved students at a much less selective institution, and ruled out the similar hypothesis that students at Yale (or similar Ivy League institutions) are unusually intellectually arrogant. Study four used four different groups of test items, in order to explore whether the specific devices covered in the previous studies were somehow responsible for the effect. It showed that they were not.

Study five involved individuals from the same participant pool as study two (i.e., Yale undergraduates) rating the explanations provided by the participants in study two both before, and after, reading the relevant expert explanations. This was to test whether the results were caused by the testing procedure, and also whether the drops in confidence recorded were merely as a result of the participants feeling 'challenged' in some way by the process. (Compare, for instance, how nervous and unsure some become while sitting examinations.) Rozenblit and Keil (2002: 535) summarise their findings as follows:

> The means of the independent ratings were much closer to the later than to the initial self-ratings of participant's knowledge. Similarly, the correlations between the independent and the self-ratings were higher for the later self-ratings. These findings support our interpretation of the drops in self-ratings shown in Studies 1–4: the participants are becoming more accurate in assessing their knowledge, not merely less optimistic or more conservative when confronted by the experimenter. . . . The IOED seems to reflect a genuine miscalibration in people's sense of how well they understand the workings of the world around them.

Rozenblit and Keil (2002) speculate about what the cause of this over-confidence might be, and isolate two key possibilities. The first is that we are prone to assume that there are essences underlying the phenomena, and that, having so assumed, we fool ourselves into thinking that we must have knowledge of these. In their own words:

> One possibility is that the illusion represents a cognitive bias like näive essentialism. . . . The essentialist bias assumes there is an essence beyond or beneath the observable. . . . An additional step is needed to conclude that an essentialist bias should lead to overconfidence about knowledge. For example, feelings of certainty about the existence of essences and hidden mechanisms may foster the conviction that one must have substantial knowledge of essences and mechanisms. Thus, an attempt to account for the strong intuition that essences and hidden mechanisms exist guides one to attribute to oneself knowledge of those hidden properties.
>
> (Rozenblit and Keil 2002: 525)

The second possibility is that the power of simple causal explanations, combined with the feeling of insight they provide us, leads us to overestimate their accuracy or depth. They write:

> Alternatively, the IOED [Illusion of Explanatory Depth] might arise from ways in which people construct skeletal but effective causal interpretations. They may wrongly attribute far too much fidelity and detail to their mental representations because the sparse renderings do have some efficacy and do provide a rush of insight . . . the very usefulness of highly sparse causal schema may create an illusion of knowing much more.
>
> (Rozenblit and Keil 2002: 525–526)

The truth of either of these alternatives would account for a human disposition to be realist. So, in a sense, which possibility obtains—assuming they are exhaustive—matters little for our purposes. Nonetheless, Rozenblit and Keil (2002) do not rest content with speculation. They continue by conducting further experiments, both to rule out further potential sources of experimental error and to better understand the overconfidence phenomenon.

2. Further Experiments: The Robustness and Scope of the Illusion

The next study, the sixth, was designed in order to determine if the participants would rate their knowledge as lower if they were given an explicit warning, at the beginning of the experiment, that they would subsequently be asked to explain the workings of, and answer questions concerning, the items they were to score their knowledge of. But even when the participants were prompted to be cautious, in this way, the effect was still present. Especially interesting is that the initial rating (T1) was no lower, as might have been expected, but that the subsequent ratings (T2–T4) were instead higher. Rozenblit and Keil (2002: 537) suggest that the warning may have affected the way that the students used the scale. They raise a possibility that seems plausible in the light of other work on cognitive dissonance, namely that the students may have striven to keep their later ratings close to their earlier ones due to a lack of a good reason for being surprised.

Rozenblit and Keil's next step was to determine whether the so-called 'illusion of explanatory depth' is a distinct phenomenon from general overconfidence (as mentioned at the beginning of Section 1). And to this end, they performed three further studies, each of which involved using a similar process to that used in the first four studies. In study seven, undergraduates were tested on ratings of self-knowledge concerning geographical facts rather than how devices operate. They used the 7-point scale to rate how well they knew the capital cities of 48 countries, one third of which are well known, one third of which are less well known, and one third of which are poorly known. Then they were asked to name the capitals of half the countries, and asked to rate their knowledge concerning those capitals again. Finally, the undergraduates were told the names of the capitals of those countries, and asked to provide a final rating of their knowledge of these. (There was therefore no analogue for the fourth phase of the first four studies, because it was not possible to ask diagnostic questions.) The results—depicted in Figure 7.3—indicated that the average overconfidence was markedly lower than in the first four studies.

But perhaps the simplicity of facts about capitals, in comparison to the accepted explanations of how devices work, was responsible for the difference? The eighth study probed this possibility by replacing facts with procedures. After providing an initial rating of their knowledge of ten procedures—e.g. how to make scrambled eggs, how to make pasta, and how to set a table—the undergraduates were asked to write down the procedure and rate their knowledge again. Crucially, they 'did not have to provide the causal connections between the steps, since the kinds of procedures queried often lacked such connections. They merely had to provide the steps in the correct sequence and in as much detail as possible' (Rozenblit and Keil 2002: 542). Then, in line with the previous study, they were given expert accounts and asked to re-rate one final time. Surprisingly, perhaps, there was no significant effect; if anything, the subjects rated their knowledge as *higher* at the later stages than they did at the earlier. (Again, see Figure 7.3.)

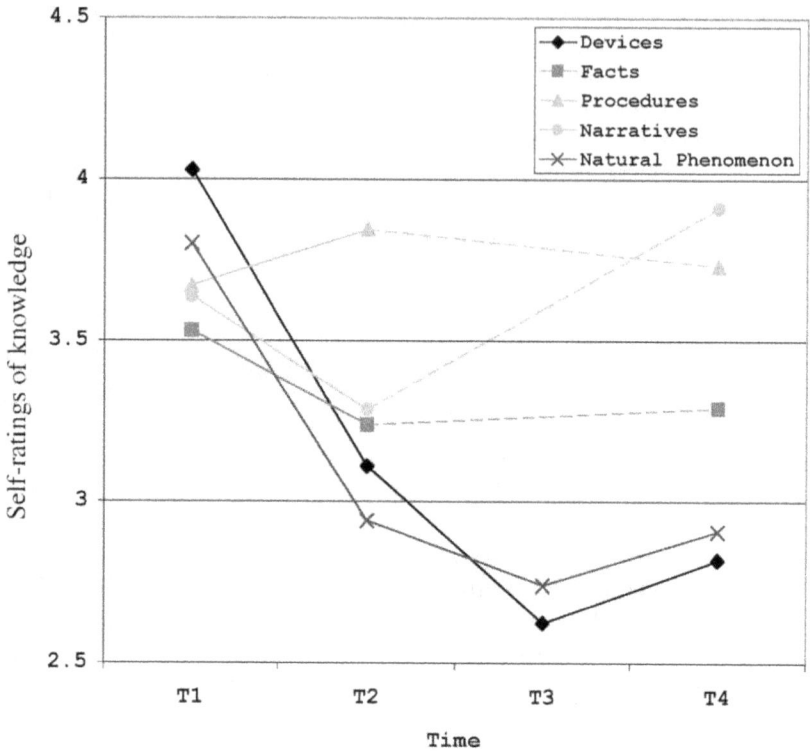

Figure 7.3 Results from Studies on Self-Knowledge Ratings in Different Domains
Reproduced with permission from Rozenblit and Keil (2002). ©2002 Leonard Rozenblit.

We now come to study nine, which was similar to study eight but concerned plot narratives rather than procedures. For this, a list of 20 popular films, devised with the aid of pilot studies, was used. The undergraduates and graduates participating were asked to declare which of the films on the list they had seen. They were then asked to rate how well they knew the plot in the first five of those films on the list. Subsequently, they re-rated their knowledge after (1) being asked to provide a plot summary, and (2) reading a plot summary by a professional reviewer. As in study eight, there was no significant drop in ratings over time. Instead, there was a small dip in ratings after (1), and a big rise after (2) such that average estimates of initial knowledge were *higher* than they were originally. Rozenblit and Keil (2002: 544) explain that this is probably due to a misalignment between the detail in the expert summary provided as an initial exemplar and the average detail of the summaries used later. They continue by suggesting that this indicates the sensitivity of the procedure.

Study ten, which is particularly interesting for our purposes, probed knowledge about natural phenomena. Rozenblit and Keil (2002: 545) introduce it as follows:

> Are devices unique? Is there something about intentionally built complex causal chains that lures us into a premature feeling of comprehension? One possibility is that the designed nature of devices is a major cause of the illusion. With designed devices, multiple levels of function encourage function/mechanism confusions. With non-living natural phenomena, functional explanations are not usually appropriate. . . . An alternative is that causally complex systems create illusions of explanatory understanding, whether intentionally designed or not, because they allow individuals to encode explanations at several distinct levels of analysis that involve stable sub-assemblies.

The structure of this study was identical to that in the first four studies (i.e., there was a diagnostic question phase, unlike in studies seven, eight, and nine). It involved undergraduates from Yale, who were initially asked to rate self-knowledge about 24 natural phenomena 'such as "how earthquakes occur," "why comets have tails," "how tides occur," and "how rainbows are formed." ' (Rozenblit and Keil 2002: 545). The later phases of the study involved five of those phenomena (selected from two possible groups of five). The results were remarkably similar to those for the first four studies:

This study is of great significance in showing that the findings of the first four studies are germane to scientific knowledge, and scientific explanation in particular. But it does not single out explanations concerning unobservables as 'special', in any sense. Hence, one might accept all the results up to this point and conclude that we are just as likely to be overconfident in our causal explanations concerning observables and our causal explanations concerning unobservables.

3. Further Experiments: The Cause of the Illusion

The penultimate study, the eleventh, was designed to rule out the possibility that the results were influenced by how socially desirable the participants perceived knowledge of the relevant domain to be. Yale undergraduates rated how embarrassing it would be, again on a 7-point scale, to have to admit ignorance concerning all the items covered in studies seven to ten. However, the results did not indicate any link between desirability and overconfidence; in fact, knowledge concerning devices was considered least desirable, although overconfidence concerning how devices work was highest on the earlier studies.[3]

The twelfth and final experiment examined whether a number of factors correlated with confidence and overconfidence. Yale undergraduates rated 40 of the devices listed in the first four studies on the following: (1) familiarity; (2) ratio of visible to hidden parts; (3) ratio of mechanical to electrical parts;[4] (4) number of parts; and (5) number of parts with known names. Measurements for (4), the number of parts, were troublesome because parts are typically decomposable into other parts—and we will touch on this issue in the next section, when we discuss unobservable parts of unobservable entities—but an instruction to think of the parts that can be acquired on disassembly 'without breaking anything' (Rozenblit and Keil 2002: 549) was introduced in order to address this problem. Also a large variation in responses, especially at the high end, was avoided by the use of a non-linear scale.

They found that neither familiarity nor the number of parts (which they refer to, somewhat misleadingly, as 'complexity') had any significant correlation with initial confidence (in study two). The other factors, however, were all significant. Yet it is not so easy to disentangle them, given a high correlation between those factors themselves: 'The ratio of visible to hidden parts was highly correlated with the mechanical:electrical dimension, with known:total parts ratio, with the raw number of parts, and with familiarity' (Rozenblit and Keil 2002: 551). The findings for relative overconfidence (over studies two to four) were similar: 'The visible/hidden parts ratio explained most of the variance in overconfidence, and adding other predictors did not significantly improve model fit. . . . [A] simple regression predicting overconfidence from the visible/hidden ratio was highly significant. . . [but] the high co-linearity between some predictor variable complicates interpretation of the step-wise regression' (Rozenblit and Keil 2002: 552). They eventually conclude:

> [T]he ratio of visible to hidden parts is the best predictor of overconfidence for an item, implicating the representational support account of the illusion.
>
> (Rozenblit and Keil 2002: 554)

In short, they take the second of the two possibilities discussed towards the end of section two to be the more likely. And they elaborate on their findings in the following passage:

> Since it is impossible in most cases to fully grasp the causal chains that are responsible for, and exhaustively explain, the world around us, we have to learn to use much sparser representations of causal relations that are good enough to give us the necessary insights: insights that go beyond associative similarity but which at the same time are not overwhelming in terms of cognitive load. It may therefore be quite adaptive to have the illusion that we know more than we do so that we settle for what is enough. The illusion might be an essential governor on our drive to search for explanatory underpinnings; it

terminates potentially inexhaustible searches for ever-deeper understanding by satiating the drive for more knowledge once some skel-
etal level of causal comprehension is reached.

(Rozenblit and Keil 2002: 558)

In essence, the finding is that we're predisposed to think we can explain
a phenomenon when we have a model that's highly predictively successful. This is a fascinating result, if we take it at face value. But let us now
consider whether we should.

4. Philosophical Objections to the Illusion of Scientific Realism Thesis

A natural way to question my appeal to these experiments is to point out
that I am a cognitive instrumentalist. So why should I take the findings of
such studies to be true, or at least approximately so? My response is twofold. First, I take the studies to concern mundane theories about observables—feelings of confidence, reports on those feelings, and so forth—for
the most part. Certainly there are no theories concerning arcane entities
such as electrons, quarks, or virtual photons being used by the empirical
psychologists in question. Some talk of beliefs (or degrees of confidence)
may creep in. But there is no need to understand these as theoretical entities such as mental representations, rather than something more homely
such as dispositions to act in particular ways, or even just actual behav-
ioural patterns. Second, and independently, scientific realists should be
inclined to accept the findings (presuming that they don't find the study
defective in method). After all, they are inclined to take what contemporary science says at face value.

One objection down. But here's another. The benchmark for expert
causal explanations, in the studies, was contemporary science. So don't
we have to assume this is correct, or at least provides the most accurate causal explanations that we have, in order for us to take the results
seriously? Irrespective of what Rozenblit and Keil intended, I think not.
Rather, a much humbler view of contemporary scientific theories, as
providing the basis for the best accounts we currently have (concerning devices and natural phenomena), suffices. Ultimately, the participants
recognised the expert accounts as better than their own when they were
shown them. And this means they would have preferred to give the same
accounts, rather than those they actually provided, if they had been
able. The realist might respond that bad accounts may often seem more
appealing than good accounts. But it would be odd for a realist to argue
this, because scientific realism appears to require, if it is to be adequately
defended, that modes of inference such as abduction—and related uses
of supra-empirical virtues in theory selection, such as simplicity, consistency, and scope—are truth conducive.[5] Of course, it would be possible to

maintain that the participants in the studies lacked the required expertise to make appropriate judgements, but this would be doubtful given the educational achievements of many of these.

A further possible objection comes from Rozenblit and Keil (2002: 554), who speculate that scientists, and others who regularly have to provide explanations, will be less prone to the illusion of explanatory depth than others. And if this is correct, then the effect will be less pronounced in the philosophical community than in the folk one. Yet even if this is accepted, it is a far stretch to say that the effect will be absent. And the specific philosophical area in which one tends to work may prove significant. For example, one feature of the *philpapers.org* survey mentioned in the introduction is a considerably lower pro-scientific-realism result for philosophers of science than for the philosophical community.

Ultimately, the issue here is the status of the so-called Expertise Defence for philosophical judgements being superior to their folk counterparts, in the present context.[6] However, although philosophers and philosophers of science are expert *explainers*, there is little reason to think they are expert *causal explainers* (unlike, say, many scientists). And the illusion of explanatory depth concerns causal explanations, not explanation in general, on the present evidence. Note also that it will not do for objectors to conclude that we should rely on scientists' judgements about scientific realism. For scientists typically lack the expertise required to appreciate the force of the arguments against scientific realism.

But one subtle rejoinder may still be available. One might say that scientific realists believe only that *scientists* are knowledgeable about causal mechanisms involving unobservables. And the studies covered in this chapter only indicate a significant probability of overconfidence in one's own knowledge. They do not indicate any overconfidence in the knowledge of others.

First, however, it is more accurate to say that many scientific realists believe themselves to be knowledgeable about many causal mechanisms involving unobservables *in virtue of receiving testimony about those mechanisms from scientists*. Indeed, almost all philosophers of science have had a science education at secondary, if not tertiary, level. Hence, the bias could enter at precisely this point. And there is evidence that a person need not think they know about something *independently of testimony* in order for the overconfidence effect to exist. Take the study involving the names of capital cities as a case in point. Capital city names seem always to be learned by testimony (construed broadly, so as to include written resources). But overconfidence is still present in the study.

Second, the illusion of explanatory depth might explain why we are inclined to think scientists are expert in this domain inappropriately. In particular, it shows that we are liable to be seduced by the following *causal* story involving unobservables. The story goes that science is so empirically successful because scientists are experts in determining how the unobservable world is. It appeals because it successfully predicts the predictive success of science.

Here is a final possible objection. Since overconfidence is correlated with the ratio of visible to hidden parts (and therefore, one might think, observable to unobservable parts) should we not conclude that it is *less* likely to be significant when dealing with theories concerning the smallest (and perhaps therefore most fundamental) unobservable things? Take quarks as a case in point. The things of which these are posited to be parts 'at the next level up' are unobservable (or 'hidden'); they have no visible parts whatsoever. Thus one might conclude that there's no reason to expect overconfidence (of the form isolated by Rozenblit and Keil) about theories concerning quarks. And the same should be true of posits such as strings.

The response to this is twofold. First, it may simply be admitted that the illusion takes part at a 'higher level' than that of quarks and strings. That is to say, it may be accepted that there is no special overconfidence when it comes to unobservable parts of unobservable things, and the unobservable mechanisms involving these. To have a strong belief in scientific realism is quite consistent, after all, with having much greater confidence in what physics says about the masses, charges, and sizes of the entities that compose the atom, than in theories concerning the composition of those entities (if any). (Talk of parts can be tricky in these contexts, but it may help to express the point as follows: the overconfidence may concern those parts of the atom that are not also *parts of particles* therein).[7] As Rozenblit and Keil (2002: 523) put it:

> Most complex artificial and natural systems are hierarchical in terms of explanations of their natures. In explaining a car one might describe the function of a unit, such as the brakes, in general terms, and then turn to describing the functions of subcomponents, such as pistons and brake pads, which in turn can be broken down even further. The iterative nature of explanations of this sort . . . may lead to an illusion of understanding when a person gains insight into a high level function and, with that rush of insight, falsely assumes an understanding of further levels down in the hierarchy of causal mechanisms.

Second the visualisability of a model may be liable to make us overconfident, in the sense that it has 'visible parts' *qua* 'parts possible to visually imagine'. For example, thinking of the atom by analogy with the solar system may tempt us into a false confidence that we know how the former works (if the existence of atoms is granted). Hence, there may be special overconfidence when it comes to such models, even when they concern only unobservable things.

Several important scientific instruments also have a high ratio of visible to 'hidden' parts. Refractor telescopes and optical microscopes are key cases in point. A sense of overconfidence that we understand how such everyday instruments work may also bolster a sense of overconfidence in the existence of light, as characterised by contemporary physical theory, and the other unobservable things posited to explain/predict their behaviour.

5. Conclusion

There is reason to suppose that we are psychologically predisposed to find scientific realism—or at least significant parts thereof—more attractive than the alternatives. So we should be especially wary, when considering the realism issue, not to let any such predisposition get the better of us. In short, to think that it's proper to adopt the 'common sense' view about scientific realism, in the absence of strong evidence to the contrary, is plausibly a mistake. Emotionally distancing oneself from the issue, and taking a cold hard look at the strength of the arguments presented here and elsewhere, is recommended.

Notes

1 See West and Stanovich (1997), Kruger and Dunning (1999), and Bjork (1998). Older studies include Adams and Adams (1960).
2 When it is plain that one cannot pass a commonly accepted and easily measured standard for being good at something, e.g. speak a language that one has never heard before, there is no overconfidence effect. See Kruger (1999). Hence Kruger and Dunning (1999: 1132) argue that: 'in order for the incompetent to overestimate themselves, they must satisfy a minimal threshold of knowledge, theory, or experience that suggests to themselves that they can generate correct answers.' This condition is satisfied in the case of philosophers of science's proclamations on science.
3 Rozenblit and Keil (2002: 547) summarise the remainder of results as: 'Movies produced the least overconfidence but the highest desirability. Natural Phenomena produced high overconfidence and intermediate desirability, while Geography Facts produced low overconfidence and intermediate desirability.'
4 The question actually asks for 'the degree to which the device operates by electrical or mechanical principles' (Rozenblit and Keil 2002: 549), but Rozenblit and Keil (2002: 548) elsewhere write of 'the number of mechanical versus electrical parts'. There is a genuine difference here, but not one that should prove significant for present purposes.
5 In particular, this is due to Duhem's thesis. If one thinks of this in a Bayesian way, for example, then the priors for theories and the auxiliary assumptions employed to derive predictions from them will be affected by their virtuosity. See Strevens (2001), for instance.
6 See, for example, Weinberg et al. (2010) and Williamson (2011). For the significance of the defence in a specific context, namely experimental semantics, see Devitt (2011a), Machery (2012), and Machery et al. (2013).
7 A more precise alternative may be to use the notion of stable subassemblies, developed by Simon (1996) and also mentioned by Rozenblit and Keil.

Appendix
What Is Scientific Realism?[1]

1. Introduction

Given the extent to which scientific realism has been discussed—for a flavour of this, consider that van Fraassen's *The Scientific Image* has been cited over 6,000 times since 1980 and that Psillos's *Scientific Realism: How Science Tracks Truth* has been cited over 1,000 times since 1999, by Google scholar's estimations—one might expect there to be considerable agreement on what, precisely, scientific realism involves. But even a perfunctory survey of the literature purporting to be on the topic dashes that hope, for as Chakravartty (2011) notes:

> It is perhaps only a slight exaggeration to say that scientific realism is characterized differently by every author who discusses it.[2]

The two influential monographs mentioned above, for instance, diverge considerably on the meaning of 'scientific realism'. On the one hand, van Fraassen (1980: 8) asserts:

> *Science aims to give us, in its theories, a literally true story of what the world is like; and acceptance of a scientific theory involves the belief that it is true.* This is the correct statement of scientific realism.

On the other hand, Psillos (1999: xix) states:

> What exactly . . . is scientific realism? I take it to incorporate three theses (or stances) . . .
>
> 1 The metaphysical stance asserts that the world has a definite and mind-independent natural-kind structure.
> 2 The semantic thesis takes scientific theories at face-value, seeing them as truth-conditioned descriptions of their intended domain, both observable and unobservable. Hence, they are capable of being true or false. . . . [I]f scientific theories are true, the unobservable entities they posit populate the world.

3 The epistemic stance regards mature and predictively successful scientific theories as well-confirmed and approximately true of the world. So, the entities posited by them, or, at any rate, entities very similar to those posited, do inhabit the world.

To reiterate, neither of these definitions of 'scientific realism' would be endorsed by the majority of experts on scientific realism. However, most experts *would* endorse definitions that bear considerable similarity to one or draw on both.[3] (In the latter case, for example, a thesis such as 'science seeks true theories' [Lyons 2005] might be added to Psillos's definition.) It is therefore apposite to consider these definitions as exemplars of two distinct, yet prominent, ways of understanding scientific realism: one *axiological*, and the other *epistemological*. In the remainder of this chapter, I'll first discuss each exemplar in turn; in doing so, I will consider how and why other definitions of the same types vary. I will also consider how the two definitional approaches are connected, and whether they can be reconciled. I will argue that they cannot be, and propose a new framework for thinking about the scientific realism debate. My view is that we should be concerned with a cluster of issues that have been discussed under the heading of 'scientific realism' in the past century or so, and which can be easily identified without appeal to any canonical positions. Terms such as 'scientific realism' and 'scientific anti-realism' may then be understood to pick out vague-boundaried resemblance classes of positions on those issues.

2. The Axiological View

Van Fraassen's view of scientific realism—and hence his alternative, constructive empiricism—involves two central notions: the aim of science and acceptance of science (and more particularly its content). Let's take each in turn, and then consider how they are connected.

The aim of science comes first. A good way to begin to discuss this is with the admission of Popper (1983: 132), writing long before van Fraassen: '[C]learly, different scientists have different aims, and science itself (whatever that may mean) has no aims.'[4] The truth of this partially explains, even if it doesn't fully excuse, the confusion that the talk of 'the aim of science' has caused in some philosophical quarters; confusion which is illustrated, for instance, by Sorensen's (2013: 30) misguided claim that scientific anti-realists are committed to theses such as 'the scientist merely aims at the prediction and control of the phenomena . . . scientists are indifferent to the truth'.

But if the aim of science isn't a function of the aims of scientists, then what is it (supposed to be)? Answering this question is far from easy, as the literature on the topic—see Resnik (1993), Rosen (1994), van Fraassen (1994), and Rowbottom (2014a)—illustrates. One might propose to answer, *prima facie*, by suggesting that Sorensen's 'scientists are indifferent to the truth' should be read as 'scientists *should* be indifferent to the truth', or the weaker 'scientists *may reasonably* be indifferent to the truth'. But such an approach is no good either. The characterisation of scientific realism offered by van Fraassen is not intended to be epistemological or methodological in character, or to bear on any other area where talk of what's obligated or permitted is appropriate, such as ethics. Rather, in his own words:

> Scientific realism and constructive empiricism are. [sic] as I understand them, not epistemologies but views of what science is. Both views characterize science as an activity with an aim—a point, a criterion of success—and construe (unqualified) acceptance of science as involving the belief that science meets that criterion.
>
> (van Fraassen 1998: 213)

Yet if scientific realism were a view of what science is, we would expect, given its popularity, for it to feature prominently in the literature on the demarcation problem. It does not.[5] Moreover, the very idea that scientific realism *centrally* concerns a thesis about the *point* of science, or what counts as success in science, is eccentric. A comparison between science and dowsing (or 'water-witching') illustrates this eccentricity. It's uncontroversial that the point of dowsing is to find water, and that an instance of dowsing is successful if water is found. And we also know that the process often succeeds, so construed. Yet this might be true even if dowsing does not result in a higher probability of finding water than choosing a location at random, and indeed proves less efficient (in so far as considerably more time consuming) than choosing a location at random.

Moreover, it seems that the *attitudes of people towards* dowsing determine the point of the exercise, construed as a process, and also what counts as success in doing it. Views on whether the process is worthwhile *depend on* views about the point of the exercise. For instance, if one thought the point of dowsing were to find *something* of interest to the dowser under the ground, then one might think dowsing worthwhile even were it to transpire that it is not a reliable means of detecting the presence of water.

I contend that something similar is true of science, which is somewhat more complex in so far as it involves many different kinds of practice (and being a scientist doesn't require being involved in, or even competent in, the full range of possible scientific activities).[6] One can learn how to perform various scientific tasks, and perform them well, without any explicit or implicit reference to an ultimate or central 'point' of the exercise— the overarching process—of which they are a part. One may focus instead on the immediate products of these tasks. (As Kuhn [1961] noted, much science education proceeds accordingly. One is judged on whether one can grasp the exemplars, employ the methods, and solve the puzzles, for instance. Whether the puzzle-solving apparatus is fit for some greater purpose is irrelevant.[7] 'What is science?' can be answered by pointing to those processes, how they interact, and so forth. And what science can achieve may be (largely or wholly) independent of what its practitioners think it can achieve, or any rather mystical 'point' of the exercise. This is a key reason why Psillos (1999: xxi) is on the right track in saying:

> It should be taken to be implicit in the realist thesis that the ampliative-abductive methods employed by scientists to arrive at their theoretical beliefs are reliable: they tend to generate approximately true beliefs and theories.

I would add that this thesis, which I will label 'methodological', should be made explicit in order to avoid confusion, and (perhaps) strengthened so that it doesn't pertain merely to methods of an 'ampliative-inductive' variety.[8] Armed with this methodological thesis, we are at an appropriate juncture to discuss acceptance.

At the heart of the concept of acceptance is a core on which the scientific realist and the anti-realist might agree; namely, that scientists sometimes adopt an attitude towards a theory such that they make:

> [A] commitment to the further confrontation of new phenomena within the framework of that theory, a commitment to a research programme, and a wager that all relevant phenomena can be accounted for without giving up that theory.
>
> (van Fraassen 1980: 88)

This is just (the pragmatic) part of 'acceptance', however; for van Fraassen, acceptance also involves belief.[9] Scientific realists think, van Fraassen alleges, that acceptance of a theory involves belief in the truth of said theory. But is this correct? I'll argue not. First, it is dubious that the pragmatic part of acceptance taken as a cluster (and hence acceptance) is significant for science. For example, one may be committed to 'a further confrontation of new phenomena within the framework' of a given theory without making any kind of 'wager that all relevant phenomena can be accounted for without giving' it up. One might merely have a 'wager' that most phenomena could be accounted for without giving it up. Or one could have no wager of that kind whatsoever, and merely be in the business of using the only theory that hasn't been refuted so far (as the best guess). In short, that's to say, although the presence of such clusters could be useful for science, this doesn't mean that they're necessary. As I've argued at length elsewhere—see Rowbottom (2011a, 2013b)—there are many ways to organise science, construed as a group endeavour, so that the core functions therein are performed satisfactorily despite the psychological facts about its participants varying considerably. This is so much so that flatly irrational individuals may contribute a great deal to the enterprise, if they have the proper roles.

Second, a scientific realist can account for a great deal of acceptance-like behaviour without thinking that it is (or should be) typically associated with belief in the truth (or, more feasibly, approximate truth) of a scientific theory. For example, one might be committed 'to further confrontation of new phenomena within the framework' of a given theory because one is convinced that the theory is false (and empirically inadequate), and wants to show that it is; and one may be similarly committed simply if one wants to discover the theory's resources (and is open-minded about what those are). One may also be committed to a research programme because one wants to see where it goes, because one dreads throwing away all the work done on it already unless absolutely necessary, or because it seems like the best programme available on the basis of its past results. And so on.

Let me make the point more bluntly. Imagine members of an alien species, for whom acceptance—or if you prefer to reserve 'acceptance' for humans, call it 'a-acceptance'—involves belief *neither in (approximate) truth nor empirical adequacy*. (This might be due to psychological constraints.[10] A-acceptance could instead involve belief in significant truth content, high problem-solving power, approximate empirical adequacy, and so on.) Would we want to say that they were incapable of doing science? Or failing that, would we want to insist that they couldn't do anything with the 'character' of science? That would be strange. For they could have institutions similar to our universities, and have theories similar to our scientific theories, arrived at by the use of similar procedures. They could also use these theories for exactly the same purposes for which we use our scientific theories: to explain the origins of the universe, to build spacecraft, and so forth.

In summary, van Fraassen's ('axiological') characterisation of realism is defective on (at least) two counts. First, it is too restrictive; it commits scientific realists to theses that they need not commit. Second, it is incomplete; it does not discuss theses to which scientific realists are typically committed. In short, van Fraassen misrepresents realism in such as way as to make it seem far less plausible than it is. Here is an example of this in action.

Van Fraassen (1998: 213) attributes the following definition to Forrest (1994):

> *scientific agnostic*: someone who believes the science s/he accepts to be empirically adequate but does not believe it to be true, nor believes it to be false.

He then offers a formulation of the opposite:

> *scientific gnostic*: someone who believes the science s/he accepts to be true.
>
> (van Fraassen 1998: 213)

He continues by declaring that:

> Scientific realists think that the scientific gnostic truly understands the character of the scientific enterprise, and that the scientific agnostic does not. The constructive empiricist thinks that the scientific gnostic may or may not understand the scientific enterprise, but that s/he adopts beliefs going beyond what science itself involves or requires for its pursuit.
>
> (van Fraassen 1998: 213–214)

The claim in the first sentence is false and uncharitable to scientific realists. That is, even if one weakens 'true', as one should, to 'approximately true' (or some near alternative). One who believes the science she accepts to be (approximately) true may do so for a variety of reasons; for example, her default attitude towards testimony from those socially recognised as experts might be to take that testimony at face value.[11] But surely realists are not committed to the claim that such a person somehow understands 'the character of the scientific enterprise'.[12] Van Fraassen's mistake appears to result from accidental inversion of a conditional. 'If you understand the character (or indeed nature) of science, *then* you will believe the scientific theories that you accept to be approximately true' *is* a claim that many scientific realists would endorse.[13] Van Fraassen's claim, on the other hand, involves swapping the antecedent with the consequent.

At the risk of overegging the pudding, here's a final *reductio* of van Fraassen's view of scientific realism. Imagine a (rather naïve) philosopher of science who thinks: (1) scientific theories should be understood literally; (2) there is a scientific method; (3) scientists invariably use this method (else what's going on isn't really science); (4) using this method guarantees that successive scientific theories become closer to the truth (construed in a correspondence sense); (5) highly predictively successful theories are approximately true; and (6) contemporary scientific theories are invariably highly predictively successful. This philosopher also believes, as a result, that (7) what contemporary science says is pretty much right, and that (8) what future science says is guaranteed to be even more right. However, he denies that (9) acceptance should be characterised in any particular way, as he thinks that's a matter for psychological investigations that haven't yet occurred.[14] (He takes psychology to be a science.) On van Fraassen's view, this philosopher is not a realist about science!

I have focused on van Fraassen's characterisation of scientific realism because of its influence. However, the idea that 'the aim of science is truth' is also present in work of several self-styled realists, most notably those influenced by Popper (or so called 'critical rationalists'). This should be of little surprise, given the quotation with which I began this section. Popper pre-empted much of what van Fraassen later said about the use of 'aim', although van Fraassen does not refer to this:

> [W]hen we speak of science, we do seem to feel . . . that there is *something characteristic of scientific activity*, and since scientific activity looks pretty much like a rational activity, and since a rational activity must have some aim, the attempt to describe the aim of science may not be entirely futile. [emphasis added]
>
> (Popper 1983: 132)

A notable philosopher working in the critical rationalist tradition is Musgrave (1988: 29), who states that: 'The aim of science, realists tell us, is to have true theories about the world, where "true" is understood in the classical correspondence sense.' However, he continues (ibid.): 'Obviously, there is more to scientific realism than a statement about the aim of science. Yet what more there is to it is a matter of some dispute among the realists themselves.' Musgrave doesn't think that acceptance has anything to do with what 'more' there is to scientific realism, however: 'If realism could explain facts about science, then it could be refuted by them too. But a philosophy of science is not a description or explanation of facts about science' (Musgrave 1988: 239). Rather, Musgrave proposes to link the axiological view and the epistemological view, to which we will soon turn our attention. His does so by suggesting that realists are committed to views about the *achievement* of the aim. His own view about this commitment is rather moderate:

> Realism . . . is the view that science aims at true theories, that sometimes it is reasonable tentatively to presume that this aim has been achieved, and that the best reason we have to presume this is novel predictive success. Thus characterised, realism explains nothing about the history of science. In particular, realism does not explain why some scientific theories have had novel predictive success.
>
> (Musgrave 1988: 234)

The problem with this view is that it seems far too weak to be of much interest, *except* in so far as it involves the aim component. That's partly because from the fact that it's *sometimes* reasonable to tentatively presume that an aim has been achieved, it doesn't follow that it's *usually* reasonable to so presume. Nor does it follow that it's ever reasonable *only* to presume that the aim has been achieved (as opposed to the contrary). (One might think it's reasonable to presume either way, say if one prefers a voluntarist epistemology.) It doesn't follow even that reasonable presumptions should be based on strong evidence. And on a related note, the best (kind of) reason for thinking something can still be a rather weak (kind of) reason. There may simply be no better (kind of) reason available. (I grant that Musgrave may have been operating with some background assumptions that make the position more interesting. But making those assumptions explicit is important.)

Nonetheless, we can see how Musgrave's approach of introducing views on the achievement of 'the aim of science' is compatible with using the methodological thesis I discussed earlier. If it is true that the methods of science 'tend to generate approximately true' beliefs and theories then it follows that doing science tends to 'achieve the aim of' generating such theories. Indeed, it's a trivial consequence. It's so trivial that it would be curious to place much emphasis on. The interesting claim is the methodological one from which the claim about achievement evidently follows.[15] To put it rather more bluntly, saying 'Science reliably does X, and achieving X is its aim' adds little of interest to 'Science reliably does X', *when 'aim' doesn't refer to the aims of the participants in the activity (as it is not supposed to in this context).* A mundane analogy may help to see the point. Consider 'Jogging reliably improves one's fitness'. If *my* aim is to improve my fitness, this is a useful thing to know; I know that jogging will help me to achieve *my* aim. It may also be of interest to empirically determine how many people jog with the aim of improving their fitness (and what their other aims in jogging are, and whether those are 'rational' in so far as jogging is a means by which to increase the probability of achieving those aims). But what's at stake in some further dispute about whether improving fitness is *characteristic* of jogging? Perhaps if one were interested in *demarcating* jogging from other activities, one might fret about this. But if one is interested mainly in how jogging works, what jogging can achieve, and how jogging technique can be improved (with reference to specific criteria like efficiency), one needn't worry about this. That is, provided it's possible—as it indeed is—to identify instances of jogging without being able to characterise or define jogging (in a philosophically serious and respectable way).

None of this is to deny that there is a worthwhile debate to be had about what the *value* of science is. Recent debate concerning scientific progress may be understood in this (non-essentialist) vein, and Chapter 1 contributes to that debate.[16] However, determining what's characteristic doesn't result in determining what's valuable. Nor is it necessary for determining what's valuable.

3. The Epistemological View

In Section 1, I presented Psillos's (1999: xix) characterisation of scientific realism, as an exemplar of the 'epistemological' class of such characterisations. And in Section 2, I also highlighted a methodological thesis that Psillos (1999: xxi) takes to be implicit in scientific realism (and which it is helpful to make explicit). I should now like to consider how other characterisations in the epistemological class vary.

As a starting point—and also as a way to remind you of Psillos's definition without summarising it or quoting from it again—it is helpful to consider a rather older, but also highly influential, definition. This is from Boyd (1980: 613):

> By 'scientific realism' philosophers ordinarily mean the doctrine that non-observational terms in scientific theories should typically be interpreted as putative referring expressions, and that when the semantics of theories is understood that way ('realistically'), scientific theories embody the sorts of propositions whose (approximate) truth can be confirmed by the ordinary experimental methods which scientists employ. There are as many possible versions of scientific realism as there are possible accounts of how 'theoretical terms' refer and of how the actual methods of science function to produce knowledge.

This passage bears on what Psillos calls the semantic and epistemic theses. Specifically, it contains one positive claim concerning the literal nature of scientific discourse about unobservable things (like Psillos's semantic thesis), and another concerning the correlation between (scientific) confirmation and approximate truth (which is part of Psillos's epistemic thesis).

Boyd (1980: 614) also explicitly endorses Psillos's metaphysical thesis (or a near equivalent): 'Reality is prior to thought . . . with respect to the correctness of theories and the appropriateness of the language in which they are expressed'.[17] (He goes further in so far as he thinks 'Reality is prior to thought . . . also with respect to the standards by which the rationality of thought is to be assessed.') And he also endorses a methodological thesis.

I delay introducing Boyd's methodological thesis, which is interestingly distinct from Psillos's, because we already have enough material to draw a significant conclusion about how 'epistemological' accounts vary, namely in so far as they involve different *qualifications*. For example, whereas Psillos's (1999: xix) semantic thesis of scientific realism states that 'The theoretical terms featuring in scientific theories have putative factual reference', Boyd's (1980: 613) equivalent only involves the more cautious claim that 'non-observational terms in scientific theories should *typically* be interpreted as putative referring expressions' [emphasis added].[18] On the other hand, Psillos's (1999: xix) epistemic thesis is more cautious than Boyd's equivalent, in so far as it only concerns 'mature' science.

Differences in qualifications can be accounted for in a relatively straightforward fashion. Such qualifications are typically introduced in order to narrow the scope of, or render more precise, the position to be articulated and discussed. For realist authors, the aim of introducing qualifications is often to modify existing statements of scientific realism so as to render them more resistant to anti-realist critiques. This comes across nicely in the following passage from Musgrave (1988: 239–240), where successively more plausible views are presented:

> It is fashionable to identify scientific realism with the view that all (or most) scientific theories are true (or nearly so), or with the view that all (or most) *current* scientific theories are true (or nearly so), or with the view that all (or most) *current* theories in the '*mature*' sciences are true (or nearly so).[19]

The view that all scientific theories are nearly true may be easily refuted by pointing to the considerable changes that have occurred in science over the past century, or even just the deep inconsistencies between competing theories at various times. For example, Thomson's model of the atom and Nagaoka's contemporary Saturnian model, discussed in Chapter 4, are different enough for it to be clear that even if the content they share is correct, at most one could be nearly true. What's more, as we saw, the latter model was of an *unstable* system, on the accepted physical laws at the time.

Musgrave mentions a variety of ways to avoid such an objection: one might adopt a variant of realism that bears only on current scientific theories, or only on theories in mature science, for example. (It is not necessary to do both, if one wants to argue that atomic theory at the turn of the twentieth-century was not mature.) Or one might take another route mentioned previously, and introduce (a high degree of) confirmation as a requirement. One might then deny that either of the aforementioned models was ever highly confirmed. Naturally, other somewhat more subtle and complicated routes are possible. For instance, one might declare that 'theory' shouldn't be understood to encompass models of the kind mentioned, or that only 'central terms' (such as 'electron') should be taken to refer successfully in the models.

The recognition that qualifications are used in the semantic and epistemic theses suggests that they have the following general form:

(Semantic) A proper subset of scientific discourse concerning unobservable entities, S, should be taken literally.[20]

(Epistemic) A proper subset of science's content, E, is approximately true (on a proper subset of theories of truth, T).[21]

Many varieties of scientific realism differ only in so far as they define S, E, or T differently. We might profitably think of them as involving different sets of sets: $\{S_1, E_1, T_1\}$, $\{S_2, E_1, T_1\}$, $\{S_1, E_2, T_1\}$, $\{S_1, E_1, T_2\}$, $\{S_2, E_2, T_1\}$, $\{S_2, E_1, T_2\}$, $\{S_1, E_2, T_2\}$, $\{S_2, E_2, T_2\}$, and so forth. We can then consider relations between such sets of sets—and, if desired, their element sets—such as similarity.

Clearly, we need to say something more about the relevant sets to get to what realism is. That is, even assuming that each variety of realism involves an identical realist metaphysical thesis like 'reality is prior to thought' (and is otherwise 'filled in' in the same way, e.g. to include the view that all scientific discourse concerning observable things should be taken literally). For as it stands, S or E might even be taken to be the empty set!

So how to move from the above to a (partial) characterisation of scientific realism? To some extent, it's helpful to think in terms of *a spectrum of positions*, ranging from complete (semantic and epistemic) realism to complete absence of (semantic and epistemic) realism. That is, in so far as we can use measures on, and assign rankings on the basis of, the sets. It's fruitful to consider S and E in turn, in the first instance. (Leave T fixed as a single member set, containing the correspondence theory of truth, for the time being.) One might think as follows. First, the size of S, relative to the set of all scientific discourse concerning unobservable entities, is a rough indicator of the strength of realism in the semantic dimension. Second, the size of E, relative to the set of all science's content (including past theories and models), is a rough indicator of the strength of realism in the epistemic dimension. (All such sets are finite.) And although this form of measurement is rather crude (and awkward), comparisons between set pairs will sometimes, at least, give clear rankings: if E_2 is a proper subset of E_1, then a version of (epistemic) employing E_1 is more realist than a version containing E_2, for example. In other cases, say where there is little or no overlap between the sets, comparisons will be fraught with difficulty. Yet this is as it should be. For example, entity realism and epistemic structural realism—discussed in Chapter 6—are evidently each less realist than scientific realism (of, say, the form endorsed by Psillos) in the epistemic dimension. Nevertheless, it's unclear which is more realist than the other.

It is also worth noting that the prior analysis provides a perspicuous way of characterising *the core of the scientific realism debate* (or more precisely, when the metaphysical thesis is assumed, the key elements thereof save the methodological one). Said debate involves tackling the following questions: what is S?; what is E?; and (to a lesser extent, in so far as there is more consensus and are fewer options) what is T? Strictly speaking, since absence of realist commitment doesn't imply anti-realist commitment—one might simply be agnostic—one should also consider sets S- and E-, which feature in two 'mirror' negative versions of (semantic) and (epistemic), as follows:

(Semantic-) A proper subset of scientific discourse concerning unobservable entities, S-, should not be taken literally.

(Epistemic-) A proper subset of science's content, E-, is not approximately true (on a proper subset of theories of truth, T).

So the debate also involves answering: what is S-?; and what is E-? In short, it concerns how to *partition* the space of discourse and the space of content into these sets.

A brief word about T is in order at this juncture. It's uncontroversial that some theories of truth are potential members, and others are not. For example, the correspondence view is, whereas the pragmatic view is not. Whether deflationary views are acceptable, however, is a more controversial matter. Suffice it to say, for present purposes, that truth-makers must be objective and mind-independent entities on an admissible theory of truth; as Psillos (1999: xxi) puts it: 'truth is a non-epistemic concept . . . assertions have truth-makers . . . these truth-makers hinge ultimately upon what the world is like.' Thus, one might fail to be a realist simply by failing to adopt an appropriate theory of truth (irrespective of how one partitions on the space of discourse and content). To make this more explicit, it is possible to remove mention of theories of truth from (epistemic) and (epistemic-), and introduce (alethic) and (alethic-) statements (involving sets T and T-). This would serve to provide a more precise analysis, but at the expense of greater complexity. The simpler route was preferable, here, partly because disputes on theories of truth are infrequent in the current debate.

We have seen that the foregoing analysis provides a way of characterising the comparative strengths of at least some forms of realism (and anti-realism), and also that it provides a relatively elegant way to characterise the scientific realism debate. The analysis also avoids—and is of help in illustrating—two reasonably common pitfalls in characterising scientific realism. The first involves appeal to *arguments*, or key propositions in arguments, that those professing to be scientific realists (tend to) employ. For example, Leplin (1984: 1) includes the following items on his list of 'characteristic claims, no majority of which, even subjected to reasonable qualification, is likely to be endorsed by any avowed realist':

3. The approximate truth of a scientific theory is sufficient explanation of its predictive success.
4. The (approximate) truth of a scientific theory is the only possible explanation of its predictive success.

However, claims such as 3 and 4 are offered to *support* variants of (semantic) and (epistemic), as realists typically acknowledge; they are used in 'no miracles' style arguments *for*—or explanationist defences *of*—scientific realism. It is important to keep them in their proper place. They could be false as a group, despite science reliably producing approximately true theories (in a correspondence sense), for example. And crucially, *neither is reducible to a statement of a general form that all realists will accept.*

The second pitfall involves characterising scientific realism in terms of knowledge, or knowledge acquisition. For example, Boyd (1980: 613) writes that scientific realism involves the claim that: 'Scientific knowledge extends to both the observable and the unobservable features of the world' and Psillos (1999: xix) claims that: 'Going for realism is going for a philosophical package which includes a naturalised approach to human knowledge.'[22] Chakravartty (2011) even goes so far as to say that approaches to defining scientific realism: 'have in common . . . a commitment to the idea that our best theories have a certain epistemic status: they yield knowledge of aspects of the world, including unobservable aspects.' And no doubt claims of this kind inspired Bird (2007b) to propose the epistemic view of scientific progress discussed in Chapter 1. The temptation to connect claims about truth (or approximate truth) with knowledge is rather natural. A realist is liable to think not only that much of science is approximately true, but also that they *know* that much of science is approximately true, and therefore know a good deal about the world in so far as they are familiar with the relevant science.

It is not necessary to succumb to this temptation in order to be a realist, however, and introducing knowledge into one's characterisation of scientific realism (and the debate) is problematic for two reasons. First, it serves to complicate matters unnecessarily, in so far as the extent to which one takes science to generate knowledge will depend on which theory of knowledge—and on which theories of related notions such as justification, warrant, and belief—one prefers. For process reliabilists, for instance, the presence of a reliable means of generating true statements will suffice for science to provide knowledge (perhaps with the addition of some provisos about belief formation, involving, for example, reliability of testimony). For internalists, on the other hand, access to reasons for belief is required for beliefs to be justified (and hence constitute items of knowledge). Even the simplest theories of knowledge introduce complications. Consider Sartwell's (1992), on which knowledge is merely true belief. Might science not isolate true claims, and rely for its successes on truth-like representations, which nobody believes in (or can fully appreciate)? Think, for instance, of the use of computer simulations. No-one can hold in their head all the detail of typical simulations used to forecast weather or to determine chemical reaction pathways, yet their predictive successes might rely on the accuracy of the modelling assumptions and other data therein. Note also that in some cases, parts of the data used might never have been believed in. Automated weather stations might provide data directly to the computer conducting the simulation, for example.

This brings us on to the second point, which is that the prior characterisations in terms of knowledge make some positions—including significant and influential positions in the history of philosophy of science—count as 'realist', or as having a 'realist' character, for the wrong reasons. Take Boyd's (1980: 613) claim as an exemplar: 'Scientific knowledge extends to both the observable and the unobservable features of the world'. Now consider Popper's (1972: 286) view that knowledge may: 'be contained in a book; or stored in a library; or taught in a university'. On this view, knowledge may be false, and not even approximately true (or anywhere close).[23] But clearly it can nevertheless be of 'observable and unobservable features of the world'. So the truth of Boyd's claim is admitted by Popper, but not in a way that is of relevance to the realism debate. Moreover, this oversight would not be fixed by using 'subjective knowledge' in definitions such as Boyd's. For Popper (1972: 111) also thought: 'traditional epistemology, with its concentration on . . . knowledge in the subjective sense, is irrelevant to the study of scientific knowledge'. Indeed, many a critical rationalist would deny that there is any subjective knowledge, above and beyond belief, in so far as she would deny that justification is possible; see Bartley (1984) and Rowbottom (2011d: ch. 1).

Granted, critical rationalism is now 'old hat'. But an emphasis on knowledge also rules out realist views that have been defended recently. Saatsi (In Press), for example, makes the case that minimal realism involves the view that scientific theories (probably) *latch on to the world* when they're predictively successful. This sets the stage for a discussion of the methodological element of realism, from which Saatsi's minimal realism is distilled.

4. The Methodological Component

The methodological component of scientific realism is introduced by Psillos (1999: xxi), recall, as follows: 'the ampliative-abductive methods employed by scientists to arrive at their theoretical beliefs are reliable: they tend to generate approximately true beliefs and theories.' A neat way to understand the core of this claim is in terms of the probability calculus, under a world-based interpretation of probability (such as a propensity view).[24] Let t represent a theory, M denote 'was selected by scientific methods', and $\approx$ denote 'is approximately true'. (Selection may involve high confirmation values, as suggested by Psillos's version of the epistemic thesis.) Then, the methodological claim is, at the bare minimum:

$$P(\approx t, Mt) > 0.5$$

It might plausibly be somewhat stronger, namely:

$$P(\approx t, Mt) \gg 0.5$$

Indeed, it would appear to be reasonable to require Psillos to specify an interval on which he takes $P(\approx t, Mt)$ to lie.

Now let's compare this with the methodological thesis associated with scientific realism by Boyd (1980: 613–614):

> [Progress] is achieved by a process of successive approximation: typically, and over time, the operation of the scientific method results in the adoption of theories which provide increasingly accurate accounts of the causal structure of the world.[25]

This suggests a rather different thesis (which would hold on the assumption that if t provides a more accurate account of the causal structure of the world than $t1$ provides, then t is more approximately true than $t1$). Let T+ represent 'is closer to the truth than', and L+ represent 'was selected later than'. Part of Boyd's claim, at the bare minimum, is:

$$P(T+(t2, t1), Mt1 \ \& \ Mt2 \ \& \ L+(t2, t1)) > 0.5$$

As before—in this case and the following—'>' might conceivably be replaced by '>>'. One might also reasonably expect an interval to be specified for the relevant probability, although none is provided.

Boyd's methodological claim also entails, more interestingly, that:

$$(Mt1 \ \& \ Mt2 \ \& \ Mt3 \ \& \ L+(t3, t2) \ \& \ L+(t2, t1)) \rightarrow$$

$$P(T+(t3, t1)) > P(T+(t2, t1))$$

Careful analysis is required to determine which variant of the methodological thesis—Boyd's or Psillos's—is bolder. On the one hand, Psillos's variant doesn't *entail* that future theories will probably be more truth-like than earlier theories; it doesn't entail, that's to say, science's probable convergence on the truth. So at first sight, it avoids the kind of 'convergent realism' that Laudan (1981) targets. On the other hand, Boyd's variant doesn't *entail* that any isolated use of scientific method(s) will probably result in an approximately true theory. It's compatible with thinking that many—or even most—theories arrived at by the use of those methods are (probably) not approximately true.

Consider now, however, what would follow if Psillos's methodological thesis were true and reasonably believed to be true by scientists. Then scientists could legitimately use said thesis to support inferences about theories. Imagine, for example, that they were comparing a new theory (selected with scientific methods) with past theories (selected in the same way). If the new theory diverged considerably from all the past theories, then there would be a very low probability that it was approximately true, in so far as there would be an exceptionally high probability that at least one of the older theories was approximately true. Thus, the scientists would have good grounds to reject the new theory. In essence, their belief in the reliability of their methods would lead them to think (it significantly more probable) that they had failed in one (recent) case, rather than repeatedly.

If approximately true theories all share significant content in common, moreover, then it follows from Psillos's thesis that considerable continuity in scientific theories is much more probable than not, over extended periods of time (albeit not continuously). Thus, it appears plausible that Psillos's thesis is stronger than Boyd's. Both are committed, to 'a [convergent] form of realism' involving 'variants of the following claims', among others:

> [Part of] R1) . . . more recent theories are closer to the truth than older theories in the same domain . . .

> R3) Successive theories in any mature science will be such that they 'preserve' the theoretical relations and apparent referents of earlier theories.

> (Laudan 1981: 20–21)

To be specific, Boyd and Psillos are committed to weaker variants of R1 and R3 involving the introduction of 'typically' or 'reliably', and hence (world-based) probability claims.

Let's now try to generalise rather more. Realists tend to think scientific methods are reliable means by which to achieve/select truth-like, or to move closer towards achieving/selecting truth-like, theories. But on the reasonable assumption that those methods involve selecting theories on the basis of virtues that they display (perhaps relative to competing theories)—predictive or explanatory power, for example—then the *underlying* claim involves linking said virtues to truthlikeness.

That's to say, there are (at least) two general forms for the methodological theses advocated by realists (where 'C' stands for 'comparative'):

(Methodological)   Scientific methods reliably (or typically) select theories or models that are virtuous.

(Methodological-C)   Scientific methods reliably (or typically) select theories or models that are more virtuous than their predecessors.

And associated with these theses are theses concerning virtues and truthlikeness (or accuracy), such as the following:

(Virtue)   Virtues are (typically) indicative of a degree, d, of truthlikeness or representational accuracy.

(Virtue-C)   If $t1$ (or $m1$) is more virtuous than $t2$ (or $m2$), then $t1$ (or $m1$) is (typically) more truth-like (or more representationally accurate) than $t2$ (or $m2$).[26]

As suggested previously, different realists will also have different views on what the relevant virtues are, how they should be ranked in order of importance (if at all), and so forth. But the details of this need not concern us here. To put it tersely, the need for theses such as (virtue), in addition to (methodological) or (methodological-C), arises because truthlikeness (or representational accuracy) cannot be directly observed, so to speak, rather than detected (or inferred). The oddity of the claim that scientific methods reliably find truth-like theories, but that those theories (typically) have nothing significant in common other than being truth-like, illustrates this. For we devise methods to enable us to select on the basis of *observable* features (whether or not we take said features to be indicative of further features). Note also that many anti-realists endorse (methodological) or (methodological-C), but not (virtue) or (virtue-C).

A significant result is that qualifications concerning the link between scientific methods and theoretical truthlikeness may arise in two distinct ways. First, one can take the methods to typically succeed (or succeed with probability P) in finding virtuous theories. Second, one can take a virtuous theory to typically be (or with probability Q be) truth-like. Lumping instances of theses such as (methodological) and (virtue) together tends to obscure this.

But are such theses necessary for scientific realism, or, failing that, central to characterising it? One might think not, at first sight, in so far as the role that they play, in combination, is to *support* theses of (epistemic) form. Nonetheless, they cannot convincingly be dispensed with. Consider, for example, a philosopher who accepts that most of the content of science is approximately true—and even that successive generations of scientific theories will be increasingly truth-like—but insists that this is a purely accidental feature of the enterprise (i.e., is a matter of mere luck). She denies that there's any link between predictive power and truthlikeness, or indeed explanatory power and truthlikeness. (Imagine, if liked, that she has a trusted source who has testified that the theories are, or will continue to be increasingly, truth-like—an alien who has seen the future of our science, or the creator of a simulation in which we dwell, or some such.) Does she count as a scientific realist? It appears not, in so far as one can imagine all the scientific realists I've cited arguing against her view of science rather vehemently. She fails to endorse any of the following aspects of the 'realist stance' that Saatsi (In Press) highlights:

> trust in the reliability of the scientific method in yielding theories that latch better and better onto the unobservable reality; trust in the corresponding objective theoretical progress of science; trust in the thesis that our best theories that make novel predictions (by and large) do so by virtue of latching onto unobservable reality.

186 *What Is Scientific Realism?*

This brings us to Saatsi's 'minimal realism', which posits a correlation between the virtue of (novel) predictive power and 'latching on to the world', or what we might call possessing a *degree* of truthlikeness or representational accuracy (rather than passing a specific threshold, e.g. in the case of 'approximate truth'). In his own words:

> [S]cience can make theoretical progress in the sense of theories latching better and better onto reality in a way that drives theories' increasing empirical adequacy and enables them to make novel predictions. Corresponding to this broader conception of theoretical progress there is a more minimal conception of realism, understood simply as a commitment to this broader kind of theoretical progress.
>
> (ibid.)

This is interesting because it *doesn't* require commitment to any E for (epistemic), although it *does* involve commitment to (methodological-C) and (virtue-C):

> [T]his kind of minimal realist commitment provides nothing like a general *recipe* that could be applied to a given current theory—e.g. the standard model of particle physics—to specify what unobservable features of the world we can claim to know.
>
> (ibid.)

Indeed, the foregoing analysis shows that there are other positions in the vicinity (and that determining which is minimal is no easy matter). For instance, one might instead couple (methodological) with (virtue-C), and appeal to the same virtue (namely, novel predictive success).

We now come to the thorny question of whether 'minimal realism' and positions in the vicinity, alluded to earlier, should be counted as forms of scientific realism. I answer in the negative, in light of the work in the tradition that I've canvassed previously, based on the centrality of theses of the form of (semantic) and (epistemic) in historical characterisations of the position. However, I take myself to have argued that positions like minimal realism are *necessary parts* of scientific realism. Hence, I don't think, for example, that one can be a scientific realist and deny (methodological) and (methodological-C) or (virtue) and (virtue-C). It follows that whether such theses hold is a significant bone of contention between realists and anti-realists (and indeed non-realists). It also follows that there are significant forms of semirealism that involve such theses: and I would call 'minimal realism' such a form of semirealism. This isn't to denigrate it in any way. It's an interesting proposal for a modest standalone position on the realism debate, which is worthy of further discussion.

5. Cognitive Instrumentalism in the Light of the Analysis

The first element of cognitive instrumentalism, argued for in the first chapter, addresses the axiological component of the realism debate (when this is understood as involving a claim about scientific progress or the value of science rather than one about 'the aim of science'). The second element of cognitive instrumentalism, argued for in the second chapter, involves taking a stance on S-, in so far as it isolates a necessary condition for discourse concerning unobservable entities to be taken literally. The third element, argued for in the third chapter (and also somewhat beforehand), involves an argument against (virtue) when d is reasonably high and (virtue-C), and thus entails a position on E and E- such that approximate truth of claims about unobservable entities shouldn't be expected (at least in connection with some of the further claims made in Chapter 6). This is the basic picture.

The foregoing treatment also makes it explicit that there are assumptions typically held by realists and anti-realists that form a background to the discussion in this book. These include at least one of (methodological) or (methodological-C)—for a set of virtues, such as accuracy, scope, and simplicity, which are largely agreed on—and that S and E contain most claims about observable entities made by corroborated/confirmed theories/models. (Some evidence for such claims has been adduced in passing, say in Chapter 4.) The modern cognitive instrumentalist, as I envisage her, will typically be committed to these claims as well as those advanced in the aforementioned chapters.

Notes

1   This chapter is based on the first half of 'Scientific Realism: What It Is, The Contemporary Debate, and New Directions', *Synthese* (2017) (https://doi.org/10.1007/s11229-017-1484-y).

2   Similar sentiments are expressed elsewhere. For example, Leplin (1984: 1) notes that 'Like the Equal Rights Movement, scientific realism is a majority position whose advocates are so divided as to appear a minority.'

3   There are some significant exceptions to this general rule. For instance, Mäki (2005: 235) denies that scientific realism should be 'taken to be an epistemological doctrine'. He writes (Mäki 2005: 236):

> I take realism to be primarily an ontological doctrine. Semantics and epistemology are important but not constitutive of the core concern of realism. On this I agree with philosophers like Michael Devitt whose formulation of scientific realism is put in more purely ontological terms: "*Scientific Realism*. Tokens of most current unobservable scientific physical types objectively exist independently of the mental" [Devitt 1991: 24].

4   Although it was not published until 1983, *Realism and the Aim of Science* is a part of the postscript to *The Logic of Scientific Discovery* that was written (and read by many in Popper's circle) in the 1950s. Elsewhere, Popper (1972: 290) instead used the notion of a regulative ideal to characterise realism:

'[the] regulative ideal of finding theories which correspond to the facts is what makes the scientific tradition a realist tradition.'

5 For example, there is not a single mention of realism (or constructive empiricism, for that matter) in the entry on 'Science and Pseudo-Science' in the *Stanford Encyclopedia of Philosophy* (Hansson 2014). There is a section on 'Criteria Based on Scientific Progress'. However, this doesn't engage with any of the recent literature on that topic, discussed in Chapter 1. 'Progress' is used in a narrower sense than most participants in the debate on scientific progress intend it.

6 Rowbottom (2014a) says more about the bearing of this variation on talk of 'the aim of science', and Rowbottom (2011c, 2013b) treat its significance with respect to scientific method. The existence of such variation is shown by a number of works in recent times, such as Galison (1997), Dupré (2001), Rowbottom (2011e), and Chang (2012).

7 Kuhn (1961: 368) allows that scientists are (typically), nevertheless, '*taught* to regard themselves as explorers and inventors who know no rules except those dictated by nature itself.' This results in 'an acquired tension . . . between professional skills on the one hand and professional ideology on the other' (ibid.: 368–369).

8 There are other methodological theses that realists might commit to as well, such as the thesis that scientists who are scientific realists (or realist in orientation) do better science than those who are not. Theses of this kind tend not to have been discussed much in journals or monographs, but are tackled by Hendry (1996), Rowbottom (2002), and Wray (2015).

9 It seems more natural to call the commitments above 'acceptance' and then discuss what kind of beliefs do, or should, accompany them. But for ease of comparison, I will follow the use that has now, alas, become standard in the literature.

10 Note that this doesn't present any obstacles to these beings doing research in a way similar to our own *at the level of the group*. For example, different members of the community may pursue different theories simultaneously.

11 'Scientific gnostics and agnostics need not be philosophers at all.' (van Fraassen 1980: 213)

12 Note also van Fraassen's slip between 'understanding the character of the enterprise' in the first sentence and 'understanding the enterprise' in the next sentence. These are different. Such imprecision (and hence lack of clarity) is, alas, characteristic of van Fraassen's discussions on this topic. For present purposes I adopt the charitable route of assuming that he means 'understanding the character' throughout.

13 I still don't think all scientific realists would agree with this claim. For one thing, 'will' should arguably be replaced with 'should'.

14 Or, if one prefers, he doesn't think there is any such thing as an aim of science, or 'success in science as such'.

15 Note that there are plausibly ways to connect the view that doing something is a reliable means by which to get closer to achieving X and the view that X is 'the aim' of doing it. See the discussion of 'the aim of science' in Rowbottom (2010b) for more on this.

16 I am now of the opinion, as I make the final revisions to the book, that there is considerable conceptual confusion underlying the use of 'scientific progress' in the literature. I will explore this in depth in Rowbottom (Forthcoming), with reference to related discourse concerning 'the aim (or task or goal) of science', 'the rationality of science', and 'the value of science', *inter alia*. My current suspicion is that what's progressive is subjective, in so far as it

depends on one's own values. On this view, the discussion in Chapter 1 may be seen as an attempt to expose what I value about science, and to encourage the reader to share those values (perhaps by revealing to the reader, via the thought experiments, that she shares them already to some extent). (Note that even on such a subjective view, something that cannot be achieved by science cannot be something to value about it. By extension, it may be possible to argue that something that science could only achieve by accident—by sheer luck—cannot be something to value about it. But I will not attempt to argue that here.)

17 The following similar thesis features in Boyd's (1983: 45) later definition of 'scientific realism': 'The reality which scientific theories describe is largely independent of our thoughts or theoretical commitments.'

18 On a later occasion, however, Boyd (1983: 45) didn't include 'typically' as a qualification: ' "Theoretical terms" in scientific theories . . . should be thought of as putatively referring expressions; scientific theories should be interpreted "realistically".'

19 He continues by noting, quite rightly, that: 'a pessimistic scientific realist might think none of these things without thereby ceasing to be a realist. A slightly more optimistic realist might tentatively accept some particular theory as true.'

20 'Entities' includes properties as well as property bearers.

21 This thesis is expressed imperfectly in so far as 'approximately' might conceivably be deleted; alternatively, one might replace 'approximately' with a variable expressing degree. Partly for reasons of economy and partly due to the current status of the debate, however, I don't include such factors in my formulation.

22 Elsewhere, Psillos (2000: 707) writes: 'The . . . *presumptuous* claim is that, although this world is independent of human cognitive activity, science can nonetheless succeed in arriving at a more or less faithful representation of it, enabling us to know the truth (or at least some truth) about it.' Again, there are two distinct claims here: arriving at a more or less faithful representation of something doesn't entail enabling us to know (some) truth about it.

23 This notion of knowledge is close to the contemporary one of 'information', at least if one does not think that information needs to be true. See Allo (2010) and Rowbottom (2014b) for more on this.

24 I assume there is some unintended imprecision in Psillos's statement: he presumably didn't intend to require that the methods *generate* the theories or beliefs, as opposed to confirm them or select them. (Confirmation or selection of generated theories will be a special case. Many generated theories will never be confirmed.)

25 Interestingly, Boyd (1980) says something similar about scientific language and scientific methods; that's to say, he takes these to improve successively too. Later, Boyd (1983: 45) also offered a weaker methodological claim:

> Scientific theories, interpreted realistically, are confirmable and in fact often confirmed as approximately true by ordinary scientific evidence interpreted in accordance with ordinary methodological standards.

26 A recent example of a version of realism subscribing to (virtue-C) is the 'relative realism' defended by Mizrahi (2013b).

Bibliography

Aaserud, F. and J. L. Heilbron. 2013. *Love, Literature, and the Quantum Atom: Niels Bohr's 1913 Trilogy Revisited*. Oxford: Oxford University Press.

Abbott, B. P. et al. 2016. 'Observation of Gravitational Waves from a Binary Black Hole Merger', *Physical Review Letters* 116, 061102.

Achinstein, P. 2001. 'Who Really Discovered the Electron?', in J. Z. Buchwald and A. Warwick (eds), *Histories of the Electron: The Birth of Microphysics*, 403–424. Cambridge, MA: MIT Press.

Adams, P. A. and J. K. Adams. 1960. 'Confidence in the Recognition and Reproduction of Words Difficult to Spell', *The American Journal of Psychology* 73, 544–552.

Ainsworth, P. M. 2010. 'What Is Ontic Structural Realism?', *Studies in History and Philosophy of Modern Physics* 41, 50–57.

Alexander, R. M. 1999. 'Engineering Approaches to Chewing and Digestion', *Science Progress* 82, 171–184.

Allo, P. 2010. 'Putting Information First: Luciano Floridi and the Philosophy of Information', *Metaphilosophy* 41, 247–254.

Anderson, J. D. 1997. *A History of Aerodynamics: And Its Impact on Flying Machines*. Cambridge: Cambridge University Press.

Andrade, E. N. da C. 1962. 'Rutherford at Manchester, 1913–1914', in J. B. Birks (ed.), *Rutherford at Manchester*, 27–42. London: Heywood & Co.

Ankeny, R. A. 2009. 'Model Organisms as Fictions', in M. Suárez (ed.), *Fictions in Science: Philosophical Essays on Modeling and Idealization*, 193–204. London: Routledge.

Arabatzis, T. 2001. 'The Zeeman Effect and the Discovery of the Electron', in J. Z. Buchwald and A. Warwick (eds), *Histories of the Electron: The Birth of Microphysics*, 171–194. Cambridge, MA: MIT Press.

Arabatzis, T. 2006. *Representing Electrons: A Biographical Approach to Theoretical Entities*. Chicago, IL: University of Chicago Press.

Armstrong, D. 1997. *A World of States of Affairs*. Cambridge: Cambridge University Press.

Armstrong, D. 1983. *What Is a Law of Nature?* Cambridge: Cambridge University Press.

Ashby, W. R. 1956. *An Introduction to Cybernetics*. London: Chapman & Hall.

Assis, A. K. T., H. Wiederkehr and G. Wolfschmidt. 2011. *Weber's Planetary Model of the Atom*. Hamburg: Tredition Science.

Ayer, A. J. 1936. *Language, Truth, and Logic.* London: Victor Gollancz.

Ayer, A. J. 1956. 'What Is a Law of Nature?', *Revue Internationale de Philosophie* 36, 144–165.

Bach, R., D. Pope, S. H. Liou and H. Batelaan. 2013. 'Controlled Double-Slit Electron Diffraction', *New Journal of Physics* 15, 033018.

Bacon, F. 1620. *Novum Organum.* Edited by P. Urbach and J. Gibson. Chicago, IL: Open Court, 1994.

Bailer-Jones, D. M. 2009. *Scientific Models in Philosophy of Science.* Pittsburgh, PA: University of Pittsburgh Press.

Baird, D. and T. Faust. 1990. 'Scientific Instruments, Scientific Progress and the Cyclotron', *British Journal for the Philosophy for Science* 41, 147–175.

Balmer, J. J. 1885. 'Notiz über die Spectrallinien des Wasserstoffs', *Annalen der Physik* 261, 80–87.

Bartley, W. W. 1984. *The Retreat to Commitment.* La Salle, IL: Open Court.

Barton, N. and L. Partridge. 2000. 'Limits to Natural Selection', *Bioessays* 22, 1075–1084.

Belot, G. In Press. 'Down to Earth Underdetermination', *Philosophy and Phenomenological Research.*

Bengson, J. and M. A. Moffett (eds). 2011. *Knowing How: Essays on Knowledge, Mind, and Action.* Oxford: Oxford University Press.

Bergmann, G. 1943. 'Outline of an Empiricist Philosophy of Physics', *American Journal of Physics* 11, 248–258 and 335–342.

Best, M. A., D. Neuhauser and L. Slavin. 2003. *Benjamin Franklin: Verification and Validation of the Scientific Process in Healthcare, as Demonstrated by the Report of the Royal Commission on Animal Magnetism and Mesmerism.* Victoria, BC: Trafford Publishing.

Bilaniuk, O. M. P., V. K. Deshpande and E. C. G. Sudarshan. 1962. ' "Meta" Relativity', *American Journal of Physics* 30, 718–723.

Bird, A. 1998. 'Dispositions and Antidotes', *Philosophical Quarterly* 48, 227–234.

Bird, A. 2000. *Thomas Kuhn.* Chesham: Acumen.

Bird, A. 2007a. *Nature's Metaphysics: Laws and Properties.* Oxford: Oxford University Press.

Bird, A. 2007b. 'What Is Scientific Progress?', *Noûs* 41, 64–89.

Bird, A. 2008. 'Scientific Progress as Accumulation of Knowledge: A Reply to Rowbottom', *Studies in History and Philosophy of Science* 39, 279–281.

Birks, J. B. (ed.) 1962. *Rutherford at Manchester.* London: Heywood & Co.

Bjork, R. A. 1998. 'Assessing Our Own Competence: Heuristics and Illusions', in D. Gopher and A. Koriat (eds), *Attention and Performance XVII: Cognitive Regulation of Performance: Interaction of Theory and Application,* 435–459. Cambridge, MA: MIT Press.

Blachowicz, J. 2009. 'How Science Textbooks Treat Scientific Method: A Philosopher's Perspective', *British Journal for the Philosophy of Science* 60, 303–344.

Blanchette, I. and K. Dunbar. 2000. 'How Analogies Are Generated: The Roles of Structural and Superficial Similarity', *Memory & Cognition* 28, 108–124.

Bohm, D. 1951. *Quantum Theory.* Englewood Cliffs, NJ: Prentice-Hall.

Bohm, D. 1952. 'A Suggested Interpretation of the Quantum Theory in Terms of "Hidden" Variables I', *Physical Review* 85, 166–179.

Bohr, N. 1913. 'On the Constitution of Atoms and Molecules', *Philosophical Magazine Series 6*, 151, 1–25.

Bohr, N. 1934. *Atomic Theory and the Description of Nature*. Cambridge: Cambridge University Press.

Bohr, N. 1958. *Atomic Theory and Human Knowledge*. London: John Wiley & Sons.

Bohr, N. 1963. *Essays 1958–1962 on Atomic Physics and Human Knowledge*. London: John Wiley & Sons.

Bokulich, A. 2009. 'Explanatory Fictions', in M. Suárez (ed.), *Fictions in Science: Philosophical Essays on Modeling and Idealization*, 91–109. London: Routledge.

Boyd, R. 1980. 'Scientific Realism and Naturalistic Epistemology', *PSA 1980 Vol. 2*, 613–662.

Boyd, R. 1983. 'On the Current Status of the Issue of Scientific Realism', *Erkenntnis* 19, 45–90.

Boyd, R. 2002. 'Scientific Realism', *Stanford Encyclopedia of Philosophy*. URL: http://plato.stanford.edu/archives/sum2002/entries/scientific-realism/

Bradie, M. 1986. 'Assessing Evolutionary Epistemology', *Biology and Philosophy* 1, 401–459.

Bressi, G., G. Carugno, R. Onofrio and G. Ruoso. 2002. 'Measurement of the Casimir Force between Parallel Metallic Surfaces', *Physical Review Letters* 88, 041804.

Brush, S. G. 1968. 'Mach and Atomism', *Synthese* 18, 192–215.

Bueno, O. 1999. 'What Is Structural Empiricism? Scientific Change in an Empiricist Setting', *Erkenntnis* 50, 59–85.

Bueno, O. 2011. 'Structural Empiricism, Again', in P. Bokulich and A. Bokulich (eds), *Scientific Structuralism*, 81–103. Dordrecht: Springer.

Butterfield, J. 2004. 'Between Laws and Models: Some Philosophical Morals of Lagrangian Mechanics', URL: http://arxiv.org/abs/physics/0409030

Butterfield, J. 2006. 'Against Pointillisme About Mechanics', *British Journal for the Philosophy of Science* 57, 709–753.

Butterfield, J. 2014. 'Our Mathematical Universe?' URL: https://arxiv.org/abs/1406.4348

Camilleri, K. and M. Schlosshauer. 2015. 'Bohr as Philosopher of Experiment: Does Decoherence Theory Challenge Bohr's Doctrine of Classical Concepts?', *Studies in History and Philosophy of Modern Physics* 49, 73–83.

Carnap, R. 1936/7, 'Testability and Meaning', *Philosophy of Science* 3, 419–471 and 4, 1–40.

Carnap, R. 1956. 'The Methodological Character of Theoretical Concepts', in H. Feigl and M. Scriven (eds), *Minnesota Studies in the Philosophy of Science Vol. 1*, 38–76. Minneapolis: University of Minnesota Press.

Carnap, R. 1974. *An Introduction to the Philosophy of Science*. New York: Basic Books.

Cartwright, D. E. 1999. *Tides: A Scientific History*. Cambridge: Cambridge University Press.

Cartwright, N. 1989. 'Capacities and Abstractions', in P. Kitcher and W. C. Salmon (eds), *Scientific Explanation*, 349–356. Minneapolis: University of Minnesota Press.

Cavendish, H. 1798. 'Experiments to Determine the Density of the Earth', *Philosophical Transactions of the Royal Society of London* 88, 469–526.

Cevolani, G. and L. Tambolo. 2013. 'Progress as Approximation to the Truth: A Defence of the Verisimilitudinarian Approach', *Erkenntnis* 78, 921–935.

Chakravartty, A. 1998. 'Semirealism', *Studies in History and Philosophy of Science* 29, 391–408.

Chakravartty, A. 2005. 'Causal Realism: Events and Processes', *Erkenntnis* 63, 7–31.

Chakravartty, A. 2007. *A Metaphysics for Scientific Realism: Knowing the Unobservable.* Cambridge: Cambridge University Press.

Chakravartty, A. 2011. 'Scientific Realism', *Stanford Encyclopedia of Philosophy.* URL: https://plato.stanford.edu/archives/sum2011/entries/scientific-realism/

Chakravartty, A. 2013. 'Realism in the Desert and in the Jungle: Reply to French, Ghins, and Psillos', *Erkenntnis* 78, 39–58.

Chalmers, D. 2002. 'Does Conceivability Entail Possibility?', in T. Gendler and J. Hawthorne (eds), *Conceivability and Possibility*, 145–200. Oxford: Oxford University Press.

Chang, H. 2012. *Is Water H2O? Evidence, Realism and Pluralism.* Dordrecht: Springer.

Choi, S. 2003. 'Improving Bird's Antidotes', *Australasian Journal of Philosophy* 81, 573–580.

Choi, S. 2006. 'The Simple vs. Reformed Conditional Analysis of Dispositions', *Synthese* 148, 369–379.

Choi, S. 2008. 'Dispositional Properties and Counterfactual Conditionals', *Mind* 117, 795–841.

Choi, S. and M. Fara. 2012. 'Dispositions', *Stanford Encyclopedia of Philosophy.* URL: http://plato.stanford.edu/archives/spr2014/entries/dispositions/

Churchland, P. 1985. 'The Ontological Status of Observables: In Praise of the Superempirical Virtues', in P. M. Churchland and C. A. Hooker (eds), *Images of Science: Essays on Realism and Empiricism, with a Reply from B. C. van Fraassen*, 35–47. Chicago, IL: University of Chicago Press.

Compton, A. H. 1921. 'The Magnetic Electron', *Journal of the Franklin Institute* 192, 145–155.

Cooper, N. 1964. 'The Aims of Science', *The Philosophical Quarterly* 57, 328–333.

Craig, W. 1953. 'On Axiomatizability Within a System', *Journal of Symbolic Logic* 18, 30–32.

Crease, R. P. 2002. 'The Most Beautiful Experiment', *Physics World* 15(9), 19–20.

Crum Brown, A. 1864. 'On the Theory of Isomeric Compounds', *Transactions of the Royal Society of Edinburgh* 23, 707–719.

Cushing, J. T. 1990. 'Is Scientific Methodology Interestingly Atemporal?', *British Journal for the Philosophy of Science* 41, 177–194.

Cushing, J. T. 1994. *Quantum Mechanics: Historical Contingency and the Copenhagen Hegemony.* Chicago, IL: University of Chicago Press.

Daumer, M., D. Dürr, S. Goldstein and N. Zanghì. 1997. 'Naive Realism About Operators', *Erkenntnis* 45, 379–397.

Dawid, R. 2013. *String Theory and the Scientific Method.* Cambridge: Cambridge University Press.

Day, C. 2012. 'The Dayside: Pretending to Hypothesize', *Physics Today* 21 June 2012. URL: http://scitation.aip.org/content/aip/magazine/physicstoday/news/the-dayside/pretendinghypothesie-a-dayside-post

De Broglie, L. 1925. *Recherches sur la Théorie des Quanta*. English translation by A. F. Kracklauer, *Annales de la Fondation Louis de Broglie*. URL: http://aflb.ensmp.fr/LDB-oeuvres/De_Broglie_Kracklauer.htm

De Finetti, B. 2008. *Philosophical Lectures on Probability*. Edited by A. Mura. Dordrecht: Springer.

Delisle, M. 1715. 'Réflexions sur l'Expérience que j'ai Rapportée à l'Académie d'un Anneau Lumineux Semblable à celui que l'on Apperçoit autour de la Lune dans les Eclipses Totales du Soleil', in *Histoire de L'Académie Royale des Sciences, Année M. DCCXV*, 166–169. Paris: Lambert & Durand (1741).

Dellsén, F. 2016. 'Scientific Progress: Knowledge Versus Understanding', *Studies in History and Philosophy of Science* 56, 72–83.

Dellsén, F. Forthcoming. 'Understanding Without Justification or Belief', *Ratio*.

De Regt, H. W. 2009. 'The Epistemic Value of Understanding', *Philosophy of Science* 76, 585–597.

De Regt, H. W. 2015. 'Scientific Understanding: Truth or Dare?', *Synthese* 192, 3781–3797.

De Regt, H. W. 2017. *Understanding Scientific Understanding*. Oxford: Oxford University Press.

De Regt, H. W. and D. Dieks. 2005. 'A Contextual Approach to Scientific Understanding', *Synthese* 144, 137–170.

De Regt, H. W. and V. Gijsbers. 2017. 'How False Theories Can Yield Genuine Understanding', in S. R. Grimm, C. Baumberger, and S. Ammon (eds), *Explaining Understanding: New Essays in Epistemology and the Philosophy of Science*, 50–75. London: Routledge.

De Solla Price, D. 1974. 'Gears from the Greeks. The Antikythera Mechanism: A Calendar Computer from ca. 80 B.C.', *Transactions of the American Philosophical Society* 64, 1–70.

Devitt, M. 1991. *Realism and Truth*. Oxford: Blackwell.

Devitt, M. 2011a. 'Experimental Semantics', *Philosophy and Phenomenological Research* 82, 418–435.

Devitt, M. 2011b. 'Are Unconceived Alternatives a Problem for Scientific Realism?', *Journal for General Philosophy of Science* 42, 285–293.

Devitt, M. 2012. 'Whither Experimental Semantics?', *Theoria* 27, 5–36.

Dingle, H. 1963. 'A Hundred Years of Spectroscopy', *British Journal for the History of Science* 1, 199–216.

Dirac, P. A. M. 1977. 'Recollections of an Exciting Era', in C. Weiner (ed.), *History of Twentieth Century Physics*, 109–146. New York: Academic Press.

Douglas, H. and P. D. Magnus. 2013. 'State of the Field: Why Novel Prediction Matters', *Studies in History and Philosophy of Science* 44, 580–589.

Dretske, F. 1977. 'Laws of Nature', *Philosophy of Science* 44, 248–268.

Duhem, P. M. M. 1954. *The Aim and Structure of Physical Theory*. Translated by P. P. Wiener. Princeton, NJ: Princeton University Press.

Dunbar, K. N. 1995. 'How Scientists Really Reason: Scientific Reasoning in Real-World Laboratories', in R. J. Sterberg and J. E. Davidson (eds), *The Nature of Insight*, 365–396. Cambridge, MA: MIT Press.

Dunbar, K. N. 2002. 'Understanding the Role of Cognition in Science: The Science as Category Framework', in P. Carruthers, S. Stich, and M. Siegal (eds), *The Cognitive Basis of Science*, 154–170. Cambridge: Cambridge University Press.

Dupré, J. 2001. *Human Nature and the Limits of Science*. Oxford: Oxford University Press.

Dürr, D., S. Goldstein and N. Zanghì. 1997. 'Bohmian Mechanics and the Meaning of the Wave Function', in R. S. Cohen, M. Horne, and J. Stachel (eds), *Experimental Metaphysics: Quantum Mechanical Studies for Abner Shimony, Vol. I*, 25–38. Dordrecht: Kluwer.

Earnshaw, S. 1842. 'On the Nature of the Molecular Forces Which Regulate the Constitution of the Luminiferous Ether', *Transactions of the Cambridge Philosophical Society* 7, 97–112.

Eckert, M. 2014. 'How Sommerfeld Extended Bohr's Model of the Atom (1913–1916)', *The European Physical Journal H* 39, 141–156.

Elgin, C. Z. 2004. 'True Enough', *Philosophical Issues* 14, 113–131.

Elgin, C. Z. 2007. 'Understanding and the Facts?', *Philosophical Studies* 132, 33–42.

Elgin, C. Z. 2008. 'Exemplification, Idealization, and Scientific Understanding', in M. Suárez (ed.), *Fictions in Science: Philosophical Essays on Modeling and Idealization*, 77–90. London: Routledge.

Elgin, C. Z. 2009. 'Is Understanding Factive?', in A. Haddock, A. Millar, and D. Pritchard (eds), *Epistemic Value*, 322–330. Oxford: Oxford University Press.

Elgin, C. Z. 2017. *True Enough*. Cambridge, MA: MIT Press.

Engel, M. 2011. 'Epistemic Luck', *Internet Encyclopedia of Philosophy*. URL: www.iep.utm.edu/epi-luck/

Esfeld, M. and V. Lam. 2008. 'Moderate Structural Realism About Space-Time', *Synthese* 160, 27–46.

Evans, E. J. 1913. 'The Spectra of Helium and Hydrogen', *Nature* 92, 5.

Evans, J. 1984. 'On the Function and Probable Origin of Ptolemy's Equant', *American Journal of Physics* 52, 1080–1089.

Eve, A. S. 1939. *Rutherford: Being the Life and Letters of the Rt. Hon. Lord Rutherford, O.M.* Cambridge: Cambridge University Press.

Everitt, C. W. F. et al. 2011. 'Gravity Probe B: Final Results of a Space Experiment to Test General Relativity', *Physical Review Letters* 106, 221101.

Falconer, I. 1987. 'Corpuscles, Electrons and Cathode Rays: J. J. Thomson and the "Discovery of the Electron" ', *British Journal for the History of Science* 20, 241–276.

Fantl, J. 2012. 'Knowledge How', *Stanford Encyclcopedia of Philosophy*. URL: http://plato.stanford.edu/archives/fall2014/entries/knowledge-how/

Faye, J. 1991. *Niels Bohr: His Heritage and Legacy. An Antirealist View of Quantum Mechanics*. Dordrecht: Kluwer.

Faye, J. 2008. 'Copenhagen Interpretation of Quantum Mechanics', *Stanford Encyclopedia of Philosophy*. URL: http://plato.stanford.edu/archives/fall2008/entries/qm-copenhagen/

Feyerabend, P. K. 1958. 'An Attempt at a Realistic Interpretation of Experience', *Proceedings of the Aristotelian Society* 58, 143–170.

Feynman, R. P. 1961. *Quantum Electrodynamics*. New York: W. A. Benjamin.

Feynman, R. P. 1965. *The Feynman Lectures on Physics*. Reading, MA: Addison-Wesley.

Fitzgerald, G. F. 1888. 'Foundations of Physical Theory: Function of Models', in J. Larmor (ed.), *The Scientific Writings of the Late George Francis Fitzgerald*, 163–169. London: Longmann, Green & Co.

Fitzgerald, G. F. 1896. 'Ostwald's Energetics', *Nature* 53, 441–442.

Folse, H. J. 1994. 'Bohr's Framework of Complementarity and the Realism Debate', in J. Faye and H. J. Folse (eds), *Niels Bohr and Contemporary Philosophy*, 119–139. Dordrecht: Kluwer.

Forrest, P. 1994. 'Why Most of Us Should Be Scientific Realists: A Reply to Van Fraassen', *Monist* 77, 47–70.

Fowler, A. 1912. 'Observations of the Principal and Other Series in the Spectrum of Hydrogen', *Monthly Notices of the Royal Astronomical Society* 73, 62–71.

Fowler, A. 1913. 'The Spectra of Helium and Hydrogen', *Nature* 92, 95–96.

Fowler, A. 1914. 'Series Lines in Spark Spectra', *Philosophical Transactions of the Royal Society A* 214, 225–266.

Frankland, E. 1870. *Lecture Notes for Chemical Students Vol. 1: Inorganic Chemistry*. London: John Van Voorst.

Freeth, T. 2009. 'Decoding an Ancient Computer', *Scientific American* 301(6), 76–83.

French, S. 1989. 'Identity and Individuality in Classical and Quantum Physics', *Australasian Journal of Philosophy* 67, 432–446.

French, S. and J. Ladyman. 2003a. 'Remodelling Structural Realism: Quantum Physics and the Metaphysics of Structure', *Synthese* 136, 31–56.

French, S. and J. Ladyman. 2003b. 'The Dissolution of Objects: Between Platonism and Phenomenalism', *Synthese* 136, 73–77.

Friedman, M. 1953. 'The Methodology of Positive Economics', in *Essays in Positive Economics*, 3–43. Chicago, IL: University of Chicago Press.

Friedman, M. 2011. 'Carnap on Theoretical Terms: Structuralism Without Metaphysics', *Synthese* 180, 249–263.

Friedrich, B. and D. Herschbach. 2003. 'Stern and Gerlach: How a Bad Cigar Helped Reorient Atomic Physics', *Physics Today* 56(12), 53.

Frigg, R. 2009. 'Models in Physics', *Routledge Encyclopedia of Philosophy*. URL: www.rep.routledge.com/articles/models-in-physics/v-1/

Frigg, R. 2010. 'Fiction and Scientific Representation', in R. Frigg and M. Hunter (eds), *Beyond Mimesis and Nominalism: Representation in Art and Science*, 97–138. Berlin and New York: Springer.

Frigg, R. and S. Hartmann. 2012. 'Models in Science', *Stanford Encyclopedia of Philosophy*. URL: https://plato.stanford.edu/archives/fall2012/entries/models-science/

Frigg, R. and I. Votsis. 2011. 'Everything You Always Wanted to Know About Structural Realism but Were Afraid to Ask', *European Journal for Philosophy of Science* 1, 227–276.

Galison, P. 1997. *Image and Logic: A Material Culture of Microphysics*. Chicago, IL: University of Chicago Press.

Gamow, G. 1939. 'Nuclear Reactions in Stellar Evolution', *Nature* 144, 575–577.

Gardner, M. R. 1979. 'Realism and Instrumentalism in 19th-Century Atomism', *Philosophy of Science* 46, 1–34.

Geiger, H. and E. Marsden. 1909. 'On a Diffuse Reflection of the α-Particles', *Proceedings of the Royal Society of London A* 82(557), 495–500.

Gelfert, A. 2003. 'Manipulative Success and the Unreal', *International Studies in the Philosophy of Science* 17, 245–263.

Gelfert, A. 2016. *How to Do Science with Models: A Philosophical Primer.* Dordrecht: Springer.

Gell-Mann, M. 1987. 'Superstring Theory', *Physica Scripta* T15, 202–209.

Gentner, D. 1983. 'Structure-Mapping: A Theoretical Framework for Analogy', *Cognitive Science* 7, 155–170.

Gentner, D. and D. R. Gentner. 1983. 'Flowing Waters or Teeming Crowds: Mental Models of Electricity', in D. Gentner and A. L. Stevens (eds), *Mental Models*, 99–129. Hillsdale, NJ: Lawrence Erlbaum.

Gentner, D., S. Brem, R. W. Ferguson, P. Wolff, A. B. Markman and K. D. Forbus. 1997. 'Analogy and Creativity in the Works of Johannes Kepler', in T. B. Ward, S. M. Smith, and J. Vaid (eds), *Creative Thought: An Investigation of Conceptual Structures and Processes*, 403–459. Washington, DC: American Psychological Association.

Gerlach, W. 1969. 'Zur Entdeckung des "Stern-Gerlach-Effektes"', *Physikalische Blätter* 25, 472.

Gillies, D. 2000. *Philosophical Theories of Probability.* London: Routledge.

Ginsparg, P. and S. Glashow. 1986. 'Desperately Seeking Superstrings?', *Physics Today* 39(5), 7–8.

Goldstein, S. 2013. 'Bohmian Mechanics', *Stanford Encyclopedia of Philosophy.* URL: http://plato.stanford.edu/archives/spr2013/entries/qm-bohm/

Gregory, R. L. 1970. *The Intelligent Eye.* London: Weidenfeld & Nicolson.

Grimm, S. R. 2011. 'Understanding', in S. Bernecker and D. Pritchard (eds), *The Routledge Companion to Understanding*, 84–94. London: Routledge.

Grimm, S. R. 2012. 'The Value of Understanding', *Philosophy Compass* 7, 103–117.

Grimm, S. R., C. Baumberger and S. Ammon (eds). 2017. *Explaining Understanding: New Essays in Epistemology and the Philosophy of Science.* London: Routledge.

Hacking, I. 1981. 'Do We See Through a Microscope?', *Pacific Philosophical Quarterly* 62, 305–322.

Hacking, I. 1983. *Representing and Intervening: Introductory Topics in the Philosophy of Natural Science.* Cambridge: Cambridge University Press.

Hájek, A. 1996. '"Mises Redux"—Redux: Fifteen Arguments Against Finite Frequentism', *Erkenntnis* 45, 209–227.

Hájek, A. 2009. 'Fifteen Arguments Against Hypothetical Frequentism', *Erkenntnis* 70, 211–235.

Hájek, A. Manuscript. 'Most Counterfactuals Are False.'

Hales, T. C. 2000. 'Cannonballs and Honeycombs', *Notices of the American Mathematical Society* 47, 440–449.

Hansson, S. O. 2014. 'Science and Pseudo-Science', *Stanford Encyclopedia of Philosophy.* URL: https://plato.stanford.edu/archives/spr2014/entries/pseudo-science/

Harker, D. 2008. 'On the Predilections for Predictions', *British Journal for the Philosophy of Science* 59, 429–453.

Harré, R. 1959. 'Notes on P. K. Feyerabend's Criticism of Positivism', *British Journal for the Philosophy of Science* 10, 43–48.

Harré, R. 1981. *Great Scientific Experiments: Twenty Experiments that Changed our View of the World*. 2002 Edition. New York: Dover.

Hawkins, R. J. and T. C. B. McLeish. 2004. 'Coarse Grained Model of Entropic Allostery', *Physical Review Letters 93*, 098104.

Heilbron, J. L. 1968. 'The Scattering of α and β Particles and Rutherford's atom', *Archive for History of Exact Sciences 4*, 247–307.

Heilbron, J. L. 1977. 'Lectures on the History of Atomic Physics 1900–1922', in C. Weiner (ed.), *History of Twentieth Century Physics*, 40–108. New York: Academic Press.

Heilbron, J. L. 2003. *Ernest Rutherford and the Explosion of Atoms*. Oxford: Oxford University Press.

Heilbron, J. L. and T. S. Kuhn. 1969. 'The Genesis of the Bohr Atom', *Historical Studies in the Physical Sciences 1*, vi–290.

Hempel, C. G. 1958. 'The Theoretician's Dilemma', in H. Feigl, M. Scriven and G. Maxwell (eds), *Minnesota Studies in the Philosophy of Science Vol. II*, 37–98. Minneapolis: University of Minnesota Press.

Hempel, C. G. 1965. *Aspects of Scientific Explanation: And Other Essays in the Philosophy of Science*. New York: Free Press.

Hempel, C. G. 2001. *The Philosophy of Carl G. Hempel: Studies in Science, Explanation, and Rationality*. Edited by J. H. Fetzer. Oxford: Oxford University Press.

Hendry, R. F. 1996. 'Realism, History and the Quantum Theory: Philosophical and Historical Arguments for Realism as a Methodological Thesis', PhD thesis, LSE. URL: http://etheses.lse.ac.uk/1442/

Hendry, R. F. 2008. 'Two Conceptions of the Chemical Bond', *Philosophy of Science 75*, 909–920.

Hendry, R. F. 2012. 'The Chemical Bond', in A. I. Woody, R. F. Hendry and P. Needham (eds), *Philosophy of Chemistry*, 293–307. San Diego: North Holland.

Hesse, M. B. 1966. *Models and Analogies in Science*. Notre Dame, IN: University of Notre Dame Press.

Hetherington, S. 2011. *How to Know: A Practicalist Conception of Knowledge*. Oxford: Wiley-Blackwell.

Hill, C. R. 1971. *Chemical Apparatus*. Oxford: Museum of the History of Science.

Hills, A. 2016. Understanding Why. *Noûs 50*, 661–688.

Hofmann, A. W. 1862. 'On Mauve and Magenta', *Proceedings of the Royal Institution of Great Britain 3*, 468–483.

Hofmann, A. W. 1865. *Introduction to Modern Chemistry, Experimental and Theoretic: Embodying Twelve Lectures Delivered at the Royal College of Chemistry*. London: Walton & Maberley.

Hooke, R. 1665. *Micrographia*. London: J. Martyn and J. Allestry.

Huber, F. 2008. 'Milne's Argument for the Log-Ratio Measure', *Philosophy of Science 75*, 413–420.

Jammer, M. 1961. *Concepts of Mass in Classical and Modern Physics*. Cambridge, MA: Harvard University Press.

Johnston, M. 1992. 'How to Speak of the Colors', *Philosophical Studies 68*, 221–263.

Kaptchuk, T. J. 1998. 'Intentional Ignorance: A History of Blind Assessment and Placebo Controls in Medicine', *Bulletin of the History of Medicine 72*, 389–433.

Keller, E. F. 2002. *Making Sense of Life: Explaining Biological Development with Models, Metaphors, and Machines.* Cambridge, MA: Harvard University Press.

Keller, E. F. 2007. 'A Clash of Two Cultures', *Nature* 445, 603.

Kelvin [Thomson, W.] 1867. 'On Vortex Atoms', *Proceedings of the Royal Society of Edinburgh* 6, 94–105.

Kelvin [Thomson, W.] 1884. *Lectures on Molecular Dynamics, and the Wave Theory of Light.* Baltimore: Johns Hopkins University Press.

Kelvin [Thomson, W.] 1902. 'Aepinus Atomized', *Philosophical Magazine Series* 6, 3, 257–283.

Kennard, E. H. 1922. 'Moment of Momentum of Magnetic Electrons', *Physical Review* 19, 420.

Kennedy, E. S. 1966. 'Late Medieval Planetary Theory', *Isis* 57, 365–378.

Khalifa, K. 2012. 'Inaugurating Understanding or Repackaging Explanation?', *Philosophy of Science* 79, 15–37.

Khalifa, K. 2017. *Understanding, Explanation, and Scientific Knowledge*, Cambridge: Cambridge University Press.

Kippenhahn, R. and A. Weigert. 1990. *Stellar Structure and Evolution.* Heidelberg: Springer.

Kitcher, P. 1993. *The Advancement of Science: Science Without Legend, Objectivity Without Illusions.* Oxford: Oxford University Press.

Klein, P. D. 2008. 'Useful False Beliefs', in Q. Smith (ed.), *Epistemology: New Essays*, 25–61. Oxford: Oxford University Press.

Kohler, R. E. 1994. *Lords of the Fly: Drosophila Genetics and the Experimental Life.* Chicago, IL: University of Chicago Press.

Kowal, C. T. and S. Drake. 1980. 'Galileo's Observations of Neptune', *Nature* 287, 311–313.

Kragh, H. 2012. *Niels Bohr and the Quantum Atom: The Bohr Model of Atomic Structure 1913–1925.* Oxford: Oxford University Press.

Kronig, R. L. 1926. 'Spinning Electrons and the Structure of Spectra', *Nature* 117, 550.

Kronig, R. L. 1960. 'The Turning Point', in M. Fierz and V. F. Weisskopf (eds), *Theoretical Physics in the Twentieth Century: A Memorial Volume to Wolfgang Pauli*, 5–39. New York: Interscience Publishers.

Kruger, J. 1999. 'Lake Wobegon Be Done! The "Below-Average Effect" and the Egocentric Nature of Comparative Ability Judgments', *Journal of Personality and Social Psychology* 77, 221–232.

Kruger, J. and D. Dunning. 1999. 'Unskilled and Unaware of It: How Difficulties in Recognizing One's Own Incompetence Lead to Inflated Self-assessments', *Journal of Personality and Social Psychology*, 77, 1121–1134.

Kuhn, T. S. 1957. *The Copernican Revolution: Planetary Astronomy in the Development of Western Thought.* Cambridge, MA: Harvard University Press.

Kuhn, T. S. 1961. 'The Function of Dogma in Scientific Research', in A. C. Crombie (ed.), *Scientific Change*, 347–369. New York: Basic Books.

Kuhn, T. S. 1962. *The Structure of Scientific Revolutions.* Chicago, IL: University of Chicago Press.

Kuhn, T. S. 1970. 'Logic of Discovery or Psychology of Research?', in I. Lakatos and A. Musgrave (eds), *Criticism and the Growth of Knowledge*, 1–23. Cambridge: Cambridge University Press.

Kuhn, T. S. 1977. *The Essential Tension: Selected Studies in Scientific Tradition and Change*. Chicago, IL: University of Chicago Press.

Kuhn, T. S. 1996. *The Structure of Scientific Revolutions*. 3rd Edition. Chicago: University of Chicago Press.

Kvanvig, J. L. 2011. 'Millar on the Value of Knowledge', *Aristotelian Society Supplementary Volume* 85, 83–99.

Ladyman, J. 1998. 'What Is Structural Realism?', *Studies in History and Philosophy of Science* 29, 409–424.

Ladyman, J. and D. Ross. 2007. *Every Thing Must Go: Metaphysics Naturalized*. Oxford: Oxford University Press.

Lamoreaux, S. K. 1997. 'Demonstration of the Casimir Force in the 0.6 to 6 μm Range', *Physical Review Letters* 78, 5–8.

Landau, L. D. and E. M. Lifshitz. 1977. *Quantum Mechanics*. 3rd Edition. New York: Pergamon.

Larmor, J. 1900. *Aether and Matter*. Cambridge: Cambridge University Press.

Laudan, L. 1977. *Progress and its Problems: Towards a Theory of Scientific Growth*. Berkeley: University of California Press.

Laudan, L. 1981. 'A Confutation of Convergent Realism', *Philosophy of Science* 48, 19–49.

Laudan, L. 1984. *Science and Values: The Aims of Science and their Role in Scientific Debate*. Berkeley: University of California Press.

Laudan, L. 1996. *Beyond Positivism and Relativism: Theory, Method and Evidence*. Boulder: Westview Press.

Lawler, I. 2016. 'Reductionism About Understanding Why', *Proceedings of the Aristotelian Society* CXVI, 229–236.

Laymon, R. 1990. 'Computer Simulations, Idealizations and Approximations', *PSA 1990 Vol. 2*, 519–534.

Leplin, J. (ed.) 1984. *Scientific Realism*. Berkeley: University of California Press.

Lewis, D. 1997. 'Finkish Dispositions', *Philosophical Quarterly* 47, 143–158.

Lipton, P. 2009. 'Understanding Without Explanation', in H. W. de Regt, S. Leonelli and K. Eigner (eds), *Scientific Understanding: Philosophical Perspectives*, 43–63. Pittsburgh: University of Pittsburgh Press.

List, C. and C. Puppe. 2009. 'Judgement Aggregation', in P. Anand, P. Pattanaik and C. Puppe (eds), *Handbook of Rational and Social Choice*, 457–482. Oxford: Oxford University Press.

Lodge, O. J. 1892. *Modern Views of Electricity*. London: Macmillan.

Longino, H. 2002. *The Fate of Knowledge*. Princeton, NJ: Princeton University Press.

Lyon, A. 2014. 'From Kolmogorov, to Popper, to Rényi: There's No Escaping Humphreys' Paradox (When Generalized)', in A. Wilson (ed.), *Chance and Temporal Asymmetry*, 112–125. Oxford: Oxford University Press.

Lyons, T. 2005. 'Towards a Purely Axiological Scientific Realism', *Erkenntnis* 63, 167–204.

MacBride, F. 2005. 'The Particular-Universal Distinction: A Dogma of Metaphysics?', *Mind* 114, 565–614.

Mach, E. 1893. *The Science of Mechanics: A Critical and Historical Account of Its Development*. 6th Edition. La Salle, IL: Open Court, 1960.

Mach, E. 1911. *The History and Root of the Principle of Conservation of Energy*. Chicago, IL: Open Court.

Mach, E. 1976. *Knowledge and Error: Sketches on the Psychology of Enquiry.* Dordrecht: Springer.

Mach, E. 1984. *The Analysis of Sensations and the Relation of the Physical to the Psychical.* Translated by C. M. Williams. La Salle, IL: Open Court.

Mach, E. 1986. *Principles of the Theory of Heat: Historically and Critically Elucidated.* Dordrecht: Springer.

Machery, E. 2012. 'Expertise and Intuitions About Reference', *Theoria* 27, 37–54.

Machery, E., R. Mallon, S. Nichols and S. P. Stich. 2013. 'If Folk Intuitions Vary, Then What?', *Philosophy and Phenomenological Research* 86, 618–635.

Magnus, P. D. and C. Callender. 2004. 'Realist Ennui and the Base Rate Fallacy', *Philosophy of Science* 71, 320–338.

Maher, P. 1988. 'Prediction, Accommodation, and the Logic of Discovery', *PSA* 1988 Vol. 1, 273–285.

Maier, C. L. 1964. 'The Role of Spectroscopy in the Acceptance of an Internally Structured Atom, 1860–1920', Ph.D. Thesis, University of Wisconsin-Madison.

Mäki, U. 2005. 'Reglobalizing Realism by Going Local or (How) Should Our Formulations of Scientific Realism be Informed About the Sciences?', *Erkenntnis* 63, 231–251.

Maraldi, M. 1723. 'Diverses Expériences d'Optique', in *Histoire de L'Académie Royale des Sciences, Année M. DCCXXIII*, p.169. Paris: Durand (1753).

Martin, C. B. 1994. 'Dispositions and Conditionals', *Philosophical Quarterly* 44, 1–8.

Matthews, M. R. 2005. 'Idealisation and Galileo's Pendulum Discoveries: Historical, Philosophical and Pedagogical Considerations', in M. R. Matthews, C. F. Gauld and A. Stinner (eds), *The Pendulum: Scientific, Historical, Philosophical and Educational Perspectives*, 209–235. Dordrecht: Springer.

Matthews, M. R., C. F. Gauld and A. Stinner (eds). 2005. *The Pendulum: Scientific, Historical, Philosophical and Educational Perspectives.* Dordrecht: Springer.

Maxwell, G. 1962. 'The Ontological Status of Theoretical Entities', in H. Feigl and G. Maxwell (eds), *Minnesota Studies in the Philosophy of Science Vol. III*, 3–27. Minneapolis: University of Minnesota Press.

McCormmach, R. 1966. 'The Atomic Theory of John William Nicholson', *Archive for History of Exact Sciences* 3, 160–184.

McCormmach, R. 2012. *Weighing the World: The Reverend John Michell of Thornhill.* Dordrecht: Springer.

McMichael, A. 1985. 'Van Fraassen's Instrumentalism', *British Journal for the Philosophy of Science* 36, 257–272.

McMullin, E. 1984. 'A Case for Scientific Realism', in J. Leplin (ed.), *Scientific Realism*, 8–40. Berkeley, CA: University of California Press.

Mehra, J. and Rechenberg, H. 1982. *The Historical Development of Quantum Theory, Vol. 1: The Quantum Theory of Planck, Einstein, Bohr and Sommerfeld: Its Foundation and the Rise of Its Difficulties 1900–1925.* Dordrecht: Springer.

Meinel, C. 2004. 'Molecules and Croquet Balls', in S. de Chadarevian and N. Hopwood (eds), *Models: The Third Dimension of Science*, 242–275. Stanford: Stanford University Press.

Mill, J. S. 1843. *A System of Logic, Ratiocinative and Inductive.* London: Longman.

Miller, D. W. 1994. *Critical Rationalism: A Restatement and Defence*. Chicago, IL: Open Court.

Milne, P. 1996. 'Log[P(h/eb)/P(h/b)] Is the One True Measure of Confirmation', *Philosophy of Science* 63, 21–26.

Mizrahi, M. 2013a. 'What Is Scientific Progress? Lessons from Scientific Practice', *Journal for General Philosophy of Science* 44, 375–390.

Mizrahi, M. 2013b. 'The Argument from Underconsideration and Relative Realism', *International Studies in the Philosophy of Science* 27, 393–407.

Mizrahi, M. and W. Buckwalter. 2014. 'The Role of Justification in the Ordinary Concept of Scientific Progress', *Journal for General Philosophy of Science* 45, 151–166.

Moghadam, R. and C. Carter. 2006. 'The Restoration of the Phillips Machine: Pumping up the Economy', *Economic Affairs* 10, 21–27.

Mohideen, U. and R. Anushree. 1998. 'Precision Measurement of the Casimir Force from 0.1 to 0.9 μm', *Physical Review Letters* 81, 004549.

Monton, B. and C. Mohler. 2014. 'Constructive Empiricism', *Stanford Encyclopedia of Philosophy*. URL: http://plato.stanford.edu/archives/spr2014/entries/constructive-empiricism/

Morgan, M. S. and M. J. Boumans. 2004. 'Secrets Hidden by Two-Dimensionality: The Economy as a Hydraulic Machine', in S. de Chadarevian and N. Hopwood (eds), *Models: The Third Dimension of Science*, 369–401. Stanford: Stanford University Press.

Morris, K. 2012. 'A Defense of Lucky Understanding', *British Journal for the Philosophy of Science* 63, 357–371.

Moss, P. A. and L. Groom. 2001. 'Microscopy', in J. Borch, M. B. Lyne, R. E. Mark and C. C. Habeger (eds), *Handbook of Physical Testing of Paper*, 149–265. Boca Raton, FL: CRC Press.

Muller, F. A. 2007. 'Inconsistency in Classical Electrodynamics?', *Philosophy of Science* 74, 253–277.

Musgrave, A. 1974. 'Logical Versus Historical Theories of Confirmation', *British Journal for the Philosophy of Science* 25, 1–23.

Musgrave, A. 1985. 'Constructive Empiricism and Realism', in P. M. Churchland and C. A. Hooker (eds), *Images of Science: Essays on Realism and Empiricism, with a Reply from B. C. van Fraassen*, 196–208. Chicago, IL: University of Chicago Press.

Musgrave, A. 1988. 'The Ultimate Argument for Scientific Realism', in R. Nola (ed.), *Relativism and Realism in Science*, 229–252. Dordrecht: Kluwer.

Nagaoka, H. 1904. 'Kinetics of a System of Particles Illustrating the Line and the Band Spectrum and the Phenomena of Radioactivity', *Philosophical Magazine Series 6*, 7, 445–455.

Nanay, B. 2013. 'Singularist Semirealism', *British Journal for the Philosophy of Science* 64, 371–394.

Nersessian, N. J. 1988. 'Reasoning from Imagery and Analogy in Scientific Concept Formation', *PSA 1988 Vol. 1*, 41–47.

Nersessian, N. J. 2002. 'The Cognitive Basis of Model-Based Reasoning in Science', in P. Carruthers, S. Stich and M. Siegal (eds), *The Cognitive Basis of Science*, 133–153. Cambridge: Cambridge University Press.

Neugebauer, O. 1975. *A History of Ancient Mathematical Astronomy*. New York: Springer.

Newton-Smith, W. H. 1981. *The Rationality of Science*. London: Routledge.

Newton-Smith, W. H. 2000. 'Introduction', in *A Companion to the Philosophy of Science*. Oxford: Blackwell.

Nicholson, J. W. 1911. 'A Structural Theory of Chemical Elements', *Philosophical Magazine* 22, 864–889.

Niiniluoto, I. 2002. *Critical Scientific Realism*. Oxford: Oxford University Press.

Niiniluoto, I. 2011. 'Scientific Progress', *Stanford Encyclopedia of Philosophy*. URL: http://plato.stanford.edu/archives/sum2015/entries/scientific-progress/

Niiniluoto, I. 2014. 'Scientific Progress As Increasing Verisimilitude', *Studies in History and Philosophy of Science* 46, 73–77.

Okasha, S. 2006. *Evolution and the Levels of Selection*. Oxford: Oxford University Press.

Paternò, E. 1869. 'Intorno All'Azione del Percloruro di Fosforo sul Clorale', *Giornale di Scienze Naturali ed Economiche di Palermo* 5, 117–122.

Pauli, W. 1946. 'Remarks on the History of the Exclusion Principle', *Science* 22, 213–215.

Pauling, L. 1970. 'Fifty Years of Progress in Structural Chemistry and Molecular Biology', *Daedalus* 99, 988–1014.

Peters, R. D. 2005. 'The Pendulum in the 21st Century—Relic or Trendsetter', in M. R. Matthews, C. F. Gauld and A. Stinner (eds), *The Pendulum: Scientific, Historical, Philosophical and Educational Perspectives*, 19–36. Dordrecht: Springer.

Petersen, A. F. 1984. 'The Role of Problems and Problem Solving in Popper's Early Work on Psychology', *Philosophy of the Social Sciences* 14, 239–250.

Phillips, A. W. 1950. 'Mechanical Models in Economic Dynamics', *Economica* 17, 283–305.

Pickering, E. C. 1897. 'The Spectrum of ζ Puppis', *Astrophysical Journal* 5, 92–94.

Poincaré, H. 1905. *Science and Hypothesis*. London: Walter Scott.

Pojman, P. 2009. 'Ernst Mach', *Stanford Encyclopedia of Philosophy*. URL: https://plato.stanford.edu/archives/sum2009/entries/ernst-mach/

Polya, G. 1973. *Mathematics and Plausible Reasoning* (Vol. 1). Princeton, NJ: Princeton University Press.

Popper, K. R. 1959. *The Logic of Scientific Discovery*. New York: Basic Books.

Popper, K. R. 1962. *Conjectures and Refutations*. New York: Basic Books.

Popper, K. R. 1970. 'Normal Science and its Dangers', in I. Lakatos and A. Musgrave (eds), *Criticism and the Growth of Knowledge*, 51–58. Cambridge: Cambridge University Press.

Popper, K. R. 1972. *Objective Knowledge: An Evolutionary Approach*. Oxford: Oxford University Press.

Popper, K. R. 1983. *Realism and the Aim of Science*. London: Routledge.

Preston, J. M. 2003. 'Kuhn, Instrumentalism, and the Progress of Science', *Social Epistemology* 17, 259–265.

Prior, E., R. Pargetter, and F. Jackson. 1982. 'Three Theses About Dispositions', *American Philosophical Quarterly* 19, 251–257.

Psillos, S. 1999. *Scientific Realism: How Science Tracks Truth*. London: Routledge.

Psillos, S. 2000. 'The Present State of the Scientific Realism Debate', *British Journal for the Philosophy of Science* 51, 705–728.

Psillos, S. 2013. 'Semirealism *or* Neo-Aristotelianism?', *Erkenntnis* 78, 29–38.

Quinton, A. 1973. *The Nature of Things*. London: Routledge.

Ramberg, P. J. 2003. *Chemical Structure, Spatial Arrangement: The Early History of Stereochemistry, 1874–1914.* Aldershot: Ashgate.

Rancourt, B. T. Forthcoming. 'Better Understanding Through Falsehood', *Pacific Philosophical Quarterly.*

Reif-Acherman, S. 2014. 'Anders Jonas Ångström and the Foundation of Spectroscopy—Commemorative Article on the Second Centenary of his Birth', *Spectrochimica Acta Part B* 102, 12–23.

Rescher, N. 1987. *Scientific Realism: A Critical Reappraisal.* Dordrecht: D. Reidel.

Resnik, D. B. 1993. 'Do Scientific Aims Justify Methodological Rules?', *Erkenntnis* 38, 223–232.

Resnik, D. B. 1994. 'Hacking's Experimental Realism', *Canadian Journal of Philosophy* 24, 395–412.

Reutlinger, A., D. Hangleiter and S. Hartmann. Forthcoming. 'Understanding (with) Toy Models', *British Journal for the Philosophy of Science.*

Rice, C. R. 2016. 'Factive Scientific Understanding Without Accurate Representation', *Biology and Philosophy* 31, 81–102.

Rickles, D. 2014. *A Brief History of String Theory: From Dual Models to M-Theory.* Dordrecht: Springer.

Ritter, C. 2001. 'An Early History of Alexander Crum Brown's Graphical Formulas', in U. Klein (ed.), *Tools and Modes of Representation in the Laboratory Sciences*, 35–46. Dordrecht: Springer.

Ritz, W. 1908. 'On a New Law of Series Spectra', *Astrophysical Journal* 28, 237–243.

Rocke, A. J. 1983. 'Subatomic Speculations on the Origin of Structure Theory', *Ambix* 30, 1–18.

Rocke, A. J. 1984. *Chemical Atomism in the Nineteenth Century: From Dalton to Cannizzaro.* Columbus: Ohio State University Press.

Rocke, A. J. 1993. *The Quiet Revolution: Hermann Kolbe and the Science of Organic Chemistry.* Berkeley and Los Angeles: University of California Press.

Rosen, G. 1994. 'What Is Constructive Empiricism?', *Philosophical Studies* 74, 143–178.

Rosi, G., F. Sorrentino, L. Cacciapuoti, M. Prevedelli and G. M. Tino. 2014. 'Precision Measurement of the Newtonian Gravitational Constant using Cold Atoms', *Nature* 510, 518–521.

Rowbottom, D. P. 2002. 'Which Methodologies of Science are Consistent with Scientific Realism?' MA Thesis, Durham. URL: http://etheses.dur.ac.uk/3752/

Rowbottom, D. P. 2008a. 'The Big Test of Corroboration', *International Studies in the Philosophy of Science* 22, 293–302.

Rowbottom, D. P. 2008b. 'N-rays and the Semantic View of Scientific Progress', *Studies in History and Philosophy of Science* 39, 277–278.

Rowbottom, D. P. 2009. 'Models in Physics and Biology: What's the Difference?', *Foundations of Science* 14, 281–294.

Rowbottom, D. P. 2010a. 'What Scientific Progress Is Not: Against Bird's Epistemic View', *International Studies in the Philosophy of Science* 24, 241–255.

Rowbottom, D. P. 2010b. 'Evolutionary Epistemology and the Aim of Science', *Australasian Journal of Philosophy* 88, 209–225.

Rowbottom, D. P. 2011a. 'The Instrumentalist's New Clothes', *Philosophy of Science* 78(5), 1200–1211.

Rowbottom, D. P. 2011b. 'Stances and Paradigms: A Reflection', *Synthese* 178, 111–119.

Rowbottom, D. P. 2011c. 'Kuhn vs. Popper on Criticism and Dogmatism in Science: A Resolution at the Group Level', *Studies in History and Philosophy of Science* 42, 117–124.

Rowbottom, D. P. 2011d. *Popper's Critical Rationalism: A Philosophical Investigation*. London: Routledge.

Rowbottom, D. P. 2011e. 'Approximations, Idealizations and "Experiments" at the Physics-Biology Interface', *Studies in History and Philosophy of Biological and Biomedical Sciences* 42, 145–154.

Rowbottom, D. P. 2013a. 'Group Level Interpretations of Probability: New Directions', *Pacific Philosophical Quarterly* 94, 188–203.

Rowbottom, D. P. 2013b. 'Kuhn Vs. Popper on Criticism and Dogmatism in Science, Part II: Striking the Balance', *Studies in History and Philosophy of Science* 44, 161–168.

Rowbottom, D. P. 2014a. 'Aimless Science', *Synthese* 191, 1211–1221.

Rowbottom, D. P. 2014b. 'Information Versus Knowledge in Confirmation Theory', *Logique et Analyse* 226, 137–149.

Rowbottom, D. P. 2014c. 'Intuitions in Science: Thought Experiments as Argument Pumps', in A. R. Booth and D. P. Rowbottom (eds), *Intuitions*, 119–134. Oxford: Oxford University Press.

Rowbottom, D. P. 2015. *Probability*. Cambridge, MA: Polity Press.

Rowbottom, D. P. Forthcoming. *Scientific Progress*. Cambridge: Cambridge University Press.

Rowbottom, D. P. and R. M. Alexander. 2011. 'The Role of Hypotheses in Biomechanical Research', *Science in Context* 25, 247–262.

Rozenblit, L. and F. Keil. 2002. 'The Misunderstood Limits of Folk Science: An Illusion of Explanatory Depth', *Cognitive Science* 26, 521–562.

Ruse, M. 1982. 'Creation Science Is Not Science', *Science, Technology, & Human Values* 7, 72–78.

Russell, B. 1911. 'Knowledge by Acquaintance and Knowledge by Description', *Proceedings of the Aristotelian Society* 11, 108–128.

Russell, B. 1918. 'The Philosophy of Logical Atomism', *The Monist* 28, 595–527.

Russell, B. 1953. 'The Cult of Common Usage', *British Journal for the Philosophy of Science* 12, 303–308.

Rutherford, E. 1938. 'Forty Years of Physics', in J. Needham and W. Pagel (eds), *Background to Modern Science*, 47–74. Cambridge: Cambridge University Press.

Saatsi, J. In Press. 'What Is Theoretical Progress of Science?', *Synthese*.

Salmon, W. C. 1990a. 'Rationality and Objectivity in Science or Tom Kuhn Meets Tom Bayes', in C. W. Savage (ed.), *Scientific Theories*, 175–204. Minneapolis: University of Minnesota Press.

Salmon, W. C. 1990b, 'The Appraisal of Theories: Kuhn Meets Bayes', *PSA 1990 Vol. 2*, 325–332.

Sankey, H. 2000. 'Methodological Pluralism, Normative Naturalism and the Realist Aim of Science', in R. Nola and H. Sankey (eds), *After Popper, Kuhn and Feyerabend: Recent Issues in Theory of Scientific Method*, 211–229. Dordrecht: Kluwer.

Sankey, H. 2008. *Scientific Realism and the Rationality of Science*. Aldershot: Ashgate.

Sartwell, C. 1992. 'Why Knowledge Is Merely True Belief', *The Journal of Philosophy* 89, 167–180.

Schiff, L. T. 1960. 'Possible New Experimental Test of General Relativity Theory', *Physical Review Letters* 4, 215.

Schorlemmer, C. 1894. *The Rise and Development of Organic Chemistry*. London: Macmillan.

Searle, J. R. 1958. 'Proper Names', *Mind* 67, 166–173.

Segall, M. H., D. T. Campbell and M. J. Herskovits. 1963. 'Cultural Differences in the Perception of Geometric Illusions', *Science* 139, 769–771.

Serway, R. A. and J. W. Jewett. 2013. *Physics for Scientists and Engineers with Modern Physics*. 9th Edition. Boston, MA: Brooks/Cole.

Shapere, D. 1969. 'Notes Towards a Post-Positivistic Interpretation of Science', in S. Barker and P. Achinstein (eds), *The Legacy of Logical Positivism*, 115–160. Baltimore: Johns Hopkins.

Shapere, D. 1974. *Galileo: A Philosophical Study*. Chicago, IL: University of Chicago Press.

Shirley, J. W. 1983. *Thomas Harriot: A Biography*. Oxford: Oxford University Press.

Simon, H. 1996. *The Sciences of the Artificial*. Cambridge, MA: MIT Press.

Smart, J. J. C. 1968. *Between Science and Philosophy*. New York: Random House.

Smith, A. 1980. 'The Principles which Lead and Direct Philosophical Enquiry: Illustrated by the History of Astronomy', in W. P. D. Wightman and J. C. Bryce (eds), *Vol. III of the Glasgow Edition of the Works and Correspondence of Adam Smith*, 31–129. Indianapolis: Liberty Fund.

Smith, A. D. 1977. 'Dispositional Properties', *Mind* 86, 439–445.

Sober, E. 1975. *Simplicity*. Oxford: Clarendon Press.

Sober, E. 1999. 'Instrumentalism Revisited', *Crítica* 31, 3–39.

Sorensen, R. 2013. 'Veridical Idealizations', in M. Frappier, L. Meynell and J. R. Brown (eds), *Thought Experiments in Science, Philosophy, and the Arts*, 30–52. London: Routledge.

Sorensen, R. 2014. 'Novice Thought Experiments', in A. R. Booth and D. P. Rowbottom (eds), *Intuitions*, 135–147. Oxford: Oxford University Press.

Stanford, P. K. 2001. 'Refusing the Devil's Bargain: What Kind of Underdetermination Should We Take Seriously?', *Philosophy of Science* 68, S1–S12.

Stanford, P. K. 2006. *Exceeding Our Grasp: Science, History, and the Problem of Unconceived Alternatives*. Oxford: Oxford University Press.

Stanley, J. 2011. *Know How*. Oxford: Oxford University Press.

Stanley, J. and T. Williamson. 2001. 'Knowing How', *Journal of Philosophy* 98, 411–444.

Stark, J. 1913. 'Observation of the Separation of Spectral Lines by an Electric Field', *Nature* 92, 401.

Strevens, M. 2001. 'The Bayesian Treatment of Auxiliary Hypotheses', *British Journal for the Philosophy of Science* 52, 515–537.

Strevens, M. 2008. *Depth: An Account of Scientific Explanation*. Cambridge, MA: Harvard University Press.

Strevens, M. 2013. 'No Understanding Without Explanation', *Studies in History and Philosophy of Science* 44, 510–515.

Strutt, J. W. 1964. *Scientific Papers*. New York: Dover.

Stuart, M. T. 2016. 'Taming Theory with Thought Experiments: Understanding and Scientific Progress', *Studies in History and Philosophy of Science Part A* 58, 24–33.

Suárez, M. 2009. 'Scientific Fictions as Rules of Inference', in M. Suárez (ed.), *Fictions in Science: Philosophical Essays on Modeling and Idealization*, 158–178. London: Routledge.

Teller, P. 2009. 'Fictions, Fictionalization, and Truth in Science', in M. Suárez (ed.), *Fictions in Science: Philosophical Essays on Modeling and Idealization*, 235–247. London: Routledge.

Thomas, L. H. 1926. 'The Motion of the Spinning Electron', *Nature* 117, 514.

Thomson, J. J. 1899. 'On the Masses of the Ions in Gases at Low Pressures', *Philosophical Magazine Series 5*, 48, 547–567.

Thomson, J. J. 1904. 'On the Structure of the Atom: An Investigation of the Stability and Periods of Oscillation of a Number of Corpuscles Arranged at Equal Intervals around the Circumference of a Circle; with Application of the Results to the Theory of Atomic Structure', *Philosophical Magazine Series 6*, 39, 237–265.

Tomonaga, S. 1997. *The Story of Spin*. Translated by T. Oka. Chicago, IL: University of Chicago Press.

Tonomura, A., J. Endo, T. Matsuda and T. Kawasaki. 1989. 'Demonstration of Single-Electron Buildup of an Interference Pattern', *American Journal of Physics* 57, 117–120.

Tooley, M. 1977. 'The Nature of Laws', *Canadian Journal of Philosophy* 7, 667–698.

Uhlenbeck, G. F. and S. A. Goudsmit. 1925. 'Ersetzung der Hypothese vom unmechanischen Zwang durch eine Forderung bezüglich des inneren Verhaltens jedes einzelnen Elektrons', *Naturwissenschaften* 13, 953–954.

Uhlenbeck, G. F. and S. A. Goudsmit. 1926. 'Spinning Electrons and the Structure of Spectra', *Nature* 117, 264–265.

Van Dover, C. 2000. *The Ecology of Deep-sea Hydrothermal Vents*. Princeton, NJ: Princeton University Press.

Van Fraassen, B. C. 1980. *The Scientific Image*. Oxford: Oxford University Press.

Van Fraassen, B. C. 1985. 'Empiricism in the Philosophy of Science', in P. M. Churchland and C. A. Hooker (eds), *Images of Science: Essays on Realism and Empiricism, with a Reply from B. C. van Fraassen*, 245–308. Chicago, IL: University of Chicago Press.

Van Fraassen, B. C. 1994. 'Gideon Rosen on Constructive Empiricism', *Philosophical Studies* 74, 179–192.

Van Fraassen, B. C. 1998. 'The Agnostic Subtly Probabilified', *Analysis* 58, 212–220.

Van Fraassen, B. C. 2001. 'Constructive Empiricism Now', *Philosophical Studies* 106, 151–170.

Van Fraassen, B. C. 2002. *The Empirical Stance*. New Haven: Yale University Press.

Van Fraassen, B. C. 2004. 'Précis of *The Empirical Stance*', *Philosophical Studies* 121, 127–132.

Van Fraassen, B. C. 2006. 'Structure: Its Shadow and Substance', *British Journal for the Philosophy of Science* 57, 275–307.

Van Fraassen, B. C. 2007. 'From a View of Science to a New Empiricism', in B. Monton (ed.), *Images of Empiricism*, 337–384. Oxford: Oxford University Press.

Van Fraassen, B. C. 2008. *Scientific Representation: Paradoxes of Perspective*. Oxford: Oxford University Press.

Vickers, P. 2013. *Understanding Inconsistent Science*. Oxford: Oxford University Press.

von Plato, Jan. 1991. 'Boltzmann's Ergodic Hypothesis', *Archive for History of Exact Sciences* 42, 71–89.

Ward, T. B., S. M. Smith and J. Vaid (eds). 1997. *Creative Thought: An Investigation of Conceptual Structures and Processes*. Washington, DC: American Psychological Association.

Weber, 1871. 'Elektrodynamische Maassbestimmungen insbesondere über das Princip der Erhaltung der Energie', *Abhandlungen der Königlichen Sächsischen Gesellschaft der Wissenschaften zu Leipzig* 10, 1–61.

Weinberg, J. M., C. Gonnerman, C. Buckner and J. Alexander. 2010. 'Are Philosophers Expert Intuiters?', *Philosophical Psychology* 23, 331–355.

Weisberg, M. 2007. 'Three Kinds of Idealization', *Journal of Philosophy* 104, 639–659.

Weisberg, M. 2008. 'Challenges to the Structural Conception of Bonding', *Philosophy of Science* 75, 932–946.

West, R. F. and K. E. Stanovich. 1997. 'The Domain Specificity and Generality of Overconfidence: Individual Differences in Performance Estimation Bias', *Psychonomic Bulletin and Review*, 4, 387–392.

Wilkenfeld, D. A. 2013. 'Understanding as Representation Manipulability', *Synthese* 190, 997–1016.

Williams, W. S. C. 1991. *Nuclear and Particle Physics*. Oxford: Oxford University Press.

Williamson, J. 2015. 'Deliberation, Judgement and the Nature of Evidence', *Economics and Philosophy* 31, 27–65.

Williamson, T. 1994. *Vagueness*. London: Routledge.

Williamson, T. 2000. *Knowledge and Its Limits*. Oxford: Oxford University Press.

Williamson, T. 2011. 'Philosophical Expertise and the Burden of Proof', *Metaphilosophy* 42, 215–229.

Wise, M. N. 1981. 'German Concepts of Force, Energy, and the Electromagnetic Ether: 1845–1880', in G. N. Cantor and M. J. S. Hodge (eds), *Conceptions of Ether: Studies in the History of Ether Theories, 1740–1900*, 269–307. Cambridge: Cambridge University Press.

Wolf, P. and G. Petit. 1997. 'Satellite Test of Special Relativity Using the Global Positioning System', *Physical Review A* 56, 4405.

Worrall, J. 1989a. 'Fresnel, Poisson and the White Spot: The Role of Successful Predictions in the Acceptance of Scientific Theories', in D. Gooding, T. Pinch and S. Schaffer (eds), *The Uses of Experiment: Studies in the Natural Sciences*, 135–157. Cambridge: Cambridge University Press.

Worrall, J. 1989b. 'Structural Realism: The Best of Both Worlds?', *Dialectica* 43, 99–124.

Wray, K. B. 2015. 'The Methodological Defense of Realism Scrutinized', *Studies in History and Philosophy of Science* 54, 74–79.

Ylikoski, P. 2009. 'The Illusion of Depth of Understanding in Science', in H. De Regt, S. Leonelli and K. Eigner (eds), *Scientific Understanding: Philosophical Perspectives*, 100–119. University of Pittsburgh Press.

Zagzebski, L. T. 2001. 'Recovering Understanding', in M. Steup (ed.), *Knowledge, Truth, and Duty: Essays on Epistemic Justification, Responsibility, and Virtue*, 235–256. Oxford: Oxford University Press.

Zeeman, P. 1897a. 'On the Influence of Magnetism on the Nature of the Light Emitted by a Substance', *Philosophical Magazine Series 5*, 262, 226–239.

Zeeman, P. 1897b. 'Doublets and Triplets in the Spectrum Produced by External Magnetic Forces', *Philosophical Magazine Series 5*, 266, 55–60 and 268, 255–259.

Index

For Product Safety Concerns and Information please contact our EU
representative GPSR@taylorandfrancis.com
Taylor & Francis Verlag GmbH, Kaufingerstraße 24, 80331 München, Germany

www.ingramcontent.com/pod-product-compliance
Lightning Source LLC
Chambersburg PA
CBHW060552220326
41598CB00024B/3080